W0112149

Innovation und Entrepreneurship

Edited by
N. Franke, Wien, Austria
D. Harhoff, München, Germany
J. Henkel, München, Germany
C. Häussler, Passau, Germany

Innovative Konzepte und unternehmerische Leistungen sind für Wohlstand und Fortschritt von entscheidender Bedeutung. Diese Schriftenreihe vereint wissenschaftliche Arbeiten zu diesem Themenbereich. Sie beschreiben substanzielle Erkenntnisse auf hohem methodischen Niveau.

Edited by
Prof. Dr. Nikolaus Franke
Wirtschaftsuniversität Wien
Wien, Austria

Prof. Dr. Joachim Henkel
Technische Universität München
München, Germany

Prof. Dietmar Harhoff , Ph.D.
Ludwig-Maximilians-Universität
München, Germany

Prof. Dr. Carolin Häussler
Universität Passau
Passau, Germany

Fabian Gäßler

Enforcing and Trading Patents

Evidence for Europe

With a Foreword by Prof. Dietmar Harhoff, PhD

 Springer Gabler

Fabian Gäßler
Munich, Germany

Dissertation Max Planck Institute for Innovation and Competition, Munich, 2015

D19

Innovation und Entrepreneurship
ISBN 978-3-658-13374-0   ISBN 978-3-658-13375-7 (eBook)
DOI 10.1007/978-3-658-13375-7

Library of Congress Control Number: 2016936427

Springer Gabler

© Springer Fachmedien Wiesbaden 2016
This work is subject to copyright. All rights are reserved by the Publisher, whether the whole or part
of the material is concerned, specifically the rights of translation, reprinting, reuse of illustrations,
recitation, broadcasting, reproduction on microfilms or in any other physical way, and transmission
or information storage and retrieval, electronic adaptation, computer software, or by similar or
dissimilar methodology now known or hereafter developed.
The use of general descriptive names, registered names, trademarks, service marks, etc. in this
publication does not imply, even in the absence of a specific statement, that such names are exempt
from the relevant protective laws and regulations and therefore free for general use.
The publisher, the authors and the editors are safe to assume that the advice and information in this
book are believed to be true and accurate at the date of publication. Neither the publisher nor the
authors or the editors give a warranty, express or implied, with respect to the material contained
herein or for any errors or omissions that may have been made.

Printed on acid-free paper

This Springer Gabler imprint is published by Springer Nature
The registered company is Springer Fachmedien Wiesbaden GmbH

Foreword

The enforcement and trade of patents in Europe have rarely been subject to empirical analysis – a fact that can partly be attributed to the scarcity of large-scale data and the institutional fragmentation at national level. Yet, research in this field appears particularly desirable in light of the upcoming introduction of the Unified Patent Court and the recent rise of activities in the market for patents. The controversial discussions on the design of the Unified Patent Court primarily focus on two themes: the potential of forum shopping, i.e., free court selection, and the effects of bifurcation, i.e., the separate treatment of validity and infringement questions.

In his doctoral thesis Fabian Gäßler first analyzes forum shopping and bifurcation in the context of the German patent litigation system. He theoretically derives propositions on litigant behavior and tests these by exploiting a comprehensive dataset on patent infringement disputes. The results suggest that the impact of certain institutional design aspects on patent holders and alleged infringers greatly depends on their financial capabilities and legal expertise.

Fabian Gäßler then introduces a newly generated dataset covering ownership changes of European and German national patents, which will provide the foundation for various empirical analyses, due to its unique scope and quality. In a first application, he analyzes the timing of patent transfers relative to events in the patent prosecution process at the European Patent Office.

Fabian Gäßler's thesis delivers intriguing new research insights which deepen our understanding of the enforcement and trade of patents in the context of European institutions. The results presented are a highly original and important contribution to the field of innovation economics and to the analysis of litigation behavior. They have relevance for practitioners, researchers, and public authorities alike.

Prof. Dietmar Harhoff, Ph.D.

Acknowledgements

In preparing this thesis I received support from many people to whom I am grateful. First and foremost, I would like to thank my advisor Dietmar Harhoff for all his guidance throughout my dissertation. He steadily encouraged me, gave very constructive comments and showed patience with me and my work in all stages of this endeavor. I am also thankful to Tobias Kretschmer who kindly agreed to serve as second advisor for my dissertation.

Special thanks go to my coauthors, Katrin Cremers, Christian Helmers and Yassine Lefouili, without whom the larger part of Chapter 3 would not have been written. Besides our joint projects, the discussions with and the comments made by them improved the quality of the remaining chapters as well.

In the last few years I enjoyed insightful discussions with fellow students, colleagues and participants at several conferences and seminars. Among these, Georg von Graevenitz and Karin Hoisl deserve particular recognition for having guided me towards the doctorate since my time as an undergraduate student at the University of Munich.

I am grateful for the financial support I received as scholarship holder and research fellow at the Max Planck Institute for Innovation and Competition.

I would like to extend my thanks to friends and my extended family for showing interest in as well as offering distraction from my research. My girlfriend Teresa deserves special mention and appreciation for all her kindness and support.

Finally, I thank my parents for their love, encouragement and faith in me throughout my life. This thesis is dedicated to them.

Fabian Gäßler

Table of Contents

List of Figures

List of Tables

Chapter 1

Introduction

Patents are regarded as a key policy instrument to spur innovation and technological progress, based on the social bargain that inventors disclose their novel and nonobvious invention to the public in return for temporary exclusion rights to use the invention. The artificial market power a patent confers upon the inventor with inevitable loss of public welfare represents a fundamental market intervention of the state. Not surprisingly, since the beginning of modern economic thought, scholars have therefore been pondering over the costs and benefits of patents to society.[1]

While early scholars focused primarily on the fundamental issue whether intellectual property rights should exist at all, research has become more and more nuanced over the last century. Starting with the seminal work of Nordhaus (1969), who was the first to consider the length of patent protection a variable parameter in patent policy, questions on the optimal patent length and scope have become a focus of attention (e.g., Gilbert and Shapiro, 1990).

However, these models, looking at one isolated invention, have not kept up with the changing technology and patent landscape. First, inventions have become increasingly cumulative, meaning that a patent on one invention has externalities on the incentives for subsequent research (Scotchmer, 1991). Second, products have become more complex, comprising multiple components covered by patents in often fragmented ownership (Shapiro, 2001). In this context, negotiations on how innovation rents are to be divided between the different parties are necessary to avoid market failure. However, these negotiations may be complicated, because patents are not always perfectly defined property rights (Merges, 1994). In fact, patents can be subject to significant uncertainty regarding their scope and validity (Lemley and Shapiro, 2005). First, patent boundaries can be vague for new technologies or abstract inventions, such as biotechnology, business methods, and software (Bessen and Meurer, 2008). In addition,

---

[1] See Menell (2000) for a historical account of economic theories concerning patents.

patent claims may be imprecisely specified by the inventor with the intention to cover subsequent technological advancements. Second, patent validity can be uncertain, because the examination procedure is imperfect, given capacity constraints at the patent office and limited access to prior art (cf. Merges, 1999; Lemley, 2001).

One consequence of uncertain validity and scope is the likely failure of negotiations regarding the distribution of quasi-rents from patents. This has led to a rise in patent disputes concerning alleged infringement and potential invalidity that need to be litigated before court. However, the costs of engaging in litigation reduce the virtue of patents as incentives for research. The provision for effective legal enforcement without creating incentives for welfare-reducing litigation activities thus becomes an integral aspect of the optimal design of patent systems.

While there has been considerable harmonization among patent systems worldwide over the last decades, this mostly refers to patent examination and less to patent litigation (Vandermeulen, 2005). Embedded in national legislation, patent litigation systems remain highly heterogeneous with fundamental differences in terms of level and recoverability of legal costs, and the availability and promptness of remedies. There has been a long-standing theoretical debate about the optimal design of patent litigation systems addressing several of these parameters (e.g., Aoki and Hu, 1999; Ayres and Klemperer, 1999; Boyce and Hollis, 2007). In contrast, insights derived from empirical analyses are limited. This can partly be attributed to two reasons. First, most patent disputes are settled privately prior to judgment or even filing. Analyzing the mere 'tip of the iceberg' population of observable disputes, scholars often are reserved in drawing clear policy recommendations. This is especially true considering that changes in the patent litigation system can have large impact beyond the court room on patenting and innovation behavior.[2] A second impediment in the empirical analysis of patent litigation systems is the fact that data collection can be a resource-intensive task, because it frequently requires accessing local records at multiple courts. While the latter is no longer true for the U.S., where the availability of structured data from multiple sources has lead to a recent rise in patent litigation studies, it remains reality for most European jurisdictions.[3]

The lack of empirical insights on patent litigation in Europe has become particularly apparent in the ongoing debate on the design of the Unified Patent Court (UPC), which will gain

---

[2]A prime example how changes in the patent litigation system can have first orders effect is the creation of the CAFC (United States Court of Appeals for the Federal Circuit) in the U.S. in 1982. While this centralized appeals court is found to have reduced legal uncertainty to the effect of more settlements (Galasso and Schankerman, 2010), it also has triggered a pro-patent shift in the patent system (Henry and Turner, 2006) causing a surge in strategic patenting in certain industries (Hall and Ziedonis, 2001; Ziedonis, 2004).

[3]For a review of the empirical literature on legal patent enforcement, see Weatherall and Webster (2014).

Europe-wide jurisdiction over infringement and revocation cases for Unitary Patents.[4] The implementation of the UPC has in general been welcomed as a solution to the currently fragmented system, where European patents *(EP)* are granted centrally but have to be enforced on a national level (Van Pottelsberghe de la Potterie, 2015). Still, policy makers, scholars, and practitioners have been arguing over the proposed design of the UPC in terms of the applied substantive law, procedural law, and court structure (Hilty *et al.*, 2012; Ullrich, 2015). Particular points of disagreement refer to the UPC's balance of two fundamental tradeoffs in the design of patent litigation systems: centralized judicial decisionmaking versus local accessibility to court, and consolidation versus bifurcation of infringement and validity issues (Wadlow, 2015).

First, in its currently planned form, the UPC will consist of multiple entry courts spatially dispersed over Europe to provide litigants with the option to seek remedies in close proximity.[5] Despite a centralized court of appeal, this plurality of courts has raised the concerns of forum shopping, where the court selecting litigant can exploit differences in decisionmaking and case management among the courts. The evaluation of this tradeoff highly depends on the question what factors determine court selection and how prone courts are to differ in their decisionmaking.

Second, while the UPC is for the most part a consolidated litigation system where the questions on infringement and validity are answered in the same proceeding, it also allows for bifurcation. In the case of bifurcation, infringement is decided by a local or regional division, and validity is heard by the central division. By having the most competent court hearing the validity issue, bifurcation is supposed to ensure high quality judgment at the complex intersection of technology and law. However, separating the issues of validity infringement may involve the risk of temporal divergence between judgment. The current UPC design already takes these aspects partly into account.[6] Still, proponents as well as opponents of bifurcation raise their concerns to the risk of either delayed enforcement or unjust enforcement on the basis of subsequently invalidated patents.

While national paradigms and clashing political agendas arguably reinforce the debate on the rules and structure of the UPC, these two design aspects find their manifestations in national

---

[4]The regulations relevant to the Unitary Patent were adopted in 2012, whereas the agreement on the Unified Patent Court was signed on 19 February 2013. The UPC is currently scheduled to commence operations at the end of 2016. Information on the latest developments can be found at http://www.unified-patent-court.org [accessed: 22 July 2015].

[5]Namely, a central division in Paris, regional divisions in London and Munich, and a still unspecified number of local divisions.

[6]For instance, it requires the participation of technically qualified judges if both issues are heard by a local or regional division. Further, the procedural rules foresee the option to stay the infringement proceeding if validity is heard separately.

patent litigation systems as well. Moore (2001b) was the first to observe that patent litigants in the U.S. deliberate select their courts, triggering a discussion about the costs and benefits of forum shopping for plaintiff, defendant, and the judiciary. Due to the increasing complexity of technological matters, scholars also raise doubts about the abilities of judges with legal background to correctly interpret patent claims. With the implementation of specialized first instance or appellate courts, and the introduction of technical judges into established courts, there is a considerable variety in the views and policies of national legislators how to address this issue (cf. Pegram, 2000; Ann, 2011).

The following two chapters of this thesis provide each a thorough empirical analysis on one of the two illustrated aspects of patent litigation systems. For this purpose, both studies use data on German patent litigation proceedings. A closer look at Germany's patent litigation system is particularly worthwhile for two reasons. First, German regional courts hear by far the largest share of patent litigation cases in Europe (Cremers *et al.*, 2013; Graham and van Zeebroeck, 2014). Second, with several courts hearing patent infringement in first instance, and a jurisdictional separation of infringement and validity issues, the German patent litigation system shares key features with the UPC and increasingly also with the U.S. patent litigation system.[7] While the results give first and foremost insights into the mechanisms of patent litigation in Germany, they may also be relevant to a more general discussion on the design of patent litigation systems and its effect on litigant behavior.

In Chapter 2 I examine the determinants of the plaintiff's court selection in patent litigation at German regional courts between 2003 and 2008. Particularly in patent litigation, disputes frequently fulfill territorial jurisdiction requirements at multiple courts, giving the plaintiff the option to conduct forum shopping, i.e., to freely select the court of his choice. While forum shopping activities are steadily observed and reported in the empirical legal literature, the determinants of court selection have remained largely unexplored. In this chapter I provide an initial analysis of the specific determinants of plaintiffs' court selection in patent litigation.

I first theoretically depict court selection as an optimization problem of the plaintiff to derive predictions about his court preferences. I consider court- and dispute-specific factors of the proceeding, such as opportunity costs due to delay in judgment and spatial distance, and analyze how these affect the plaintiff's expected utility of going to a particular court with the endogenous chance of settlement.

---

[7]The U.S. patent litigation system is characterized by a federal court structure with multiple entry courts hearing patent cases. Further, the Inter Partes Review (IPR), introduced in September 2012 as a way of challenging validity before the U.S. Patent and Trademark Office, has to some degree introduced bifurcation (Chien and Helmers, 2015).

I then test my predictions on the plaintiff's preferences using a comprehensive dataset on patent litigation in Germany which covers all proceedings filed at the regional courts in Düsseldorf, Mannheim, and Munich between 2003 and 2008. One key variable in determining the plaintiff's opportunity costs due to delay in judgment is the ex ante expected time of proceeding, which is not observable. Here, I exploit the rigid structure of German patent proceedings and depict the time until judgment as a function of the court's caseload and the chance of two delaying events: the request for an expert opinion and the stay of the proceeding due to a parallel validity challenge.

Using binary response models, I regress both the events of an expert opinion and a stay of proceeding on case- and court-specific variables. The results indicate that judges differ in their tendency to request an expert depending on the patent's technology. This suggests that judges are specialized and gain expertise in particular technologies from exposure to prior patent cases. Furthermore, I find differences in the judges' propensity to stay the proceeding due to a parallel validity challenge.

In the final stage of the empirical analysis, I use alternative-specific conditional logit models to recover estimates of the plaintiffs' preferences for the three courts. The primary results reveal that plaintiffs consider their economic loss from delayed judgment when choosing their court. In line with the theoretical predictions, speedy enforcement is most valued, if the litigants operate in the same product market. I also find that the distance to a particular court has a significant negative effect on court selection. Especially small plaintiffs highly value local access to court. The results also contain some indications that courts do not show perfect uniformity in their decisionmaking.

Chapter 3, which is joint work with Katrin Cremers, Dietmar Harhoff, Christian Helmers, and Yassine Lefouili, analyzes the functioning of bifurcated patent litigation systems, where infringement and validity are decided independently by different courts, under special consideration of uncertainty regarding a patent's validity and scope.

We first propose a theoretical model which demonstrates that bifurcation favors the patent holder due to the possibility of patent enforcement with little delay in judgment, and the lower likelihood of facing a validity challenge. If patents serve as an incentive mechanism to encourage investments in innovation, strong rights to enforce a patent against alleged infringers may even be socially desirable. However, the model also shows that a bifurcated system distorts the patent holder's incentives when patents are uncertain in terms of validity. These distortions originate from the lower likelihood of facing a validity challenge for a potentially invalid patent and the possibility to enforce an invalid patent during the injunction gap.

We then try to empirically quantify the extent to which bifurcation deters validity challenges and creates such 'invalid but infringed' decisions. For this purpose, we use combined case level data on German infringement and invalidity proceedings for 2000 to 2008. We first show that in practice the decision on infringement is often rendered and enforced before validity has been determined. Our data reveal that 12% of infringement cases with parallel invalidity proceedings produce divergent, i.e., 'invalid but infringed', decisions. Looking at the temporal gap between the two decisions, we find that the infringement decision was on average enforceable for more than a year before the patent was invalidated. These results show that bifurcation offers scope for patent holders to temporarily enforce invalid patents.

We then show that bifurcation reduces the likelihood that an alleged infringer challenges a patent's validity in the first place. Using probit models with a parallel revocation action as dependent variable, we find that smaller and foreign firms are less likely to challenge validity when faced with infringement allegations.

In the third part of our empirical analysis, the results of a differences-in-differences estimation show that alleged infringers subject to a divergent decision file more oppositions immediately afterwards. We interpret this as evidence that firms attempt to preempt similar situations in the future by eliminating potentially threatening patents early on.

In Chapter 4 I depart from the litigation context, but stay focused on the overall question of how patent holders can benefit from their inventions and how uncertainty may have an effect on this. While the rights to exclude others from making, using, or selling the protected technology originate with the inventor, they can be transferred to and asserted by anyone who gains ownership of the patent. This provides the inventor with a way to appropriate rents from his invention independent of his ability of commercialization and enforcement. In light of this, the transfer of patents promotes the vertical and horizontal disintegration of knowledge-based industries (Hall and Ziedonis, 2001; Arora *et al.*, 2004). Apart from its effect on industry structure, the market for patents is also regarded as a means to rectify the initial mismatch between locus of idea and of best-possible utilization and to reduce hold up problems in the context of cumulative innovation (Akcigit *et al.*, 2013; Spulber, 2015).

In return, scholars have also pointed to several challenges to successful transactions in the market for patents, such as transaction costs and information asymmetries in patent quality and value. I pursue this line of research in this study and investigate the development of the market for patents in Europe and the impact of uncertainty over patent quality on the timing of patent transactions. However, analyzing the market for patents in Europe poses a considerable data challenge. First, collecting and combining data from multiple authorities is necessary due to the fragmented nature of the European patent system. Second, the raw data retrieved contain all possible registered ownership changes without differentiation, which considerably complicates the identification of market transactions.

After a brief survey on prior patent transfer studies, I illustrate the generation process of a novel dataset covering registered ownership changes of *DE* and *EP* patents between 1981 and 2013, the *MPI-IC Patent Transfers Data 2015* (MPIIC-PT2015). To ease the subsequent empirical analysis of the market for patents, I introduce a simple taxonomy of patent transfers that takes into account the relational and spatial distance between current and prior rights holders. According to this taxonomy, I then classify all patent ownership changes in the data. This classification is vital for the subsequent empirical analysis, because only a subset of all patent transfers, namely disembodied transactions at arm's length, can be considered evidence for the market for patents.

On basis of patent transfers that are identified as being arm's length, I then examine the activities in the market for patents in Europe. In a first descriptive analysis, I find that there has been a considerable increase in absolute patent transfers since the late 1990s. The transfer rate by granted patents per year remains relatively constant at 7 to 8%. While comparability remains questionable, this result suggests a significantly lower transfer rate than those found for the U.S. market for patents.

At last, I look at the timing of patent transfers relative to the patents' grant dates. Using a Cox proportional hazard model, I find that the hazard of patent transfer increases significantly after the internal communication of patent grant to the rights holder. I see these findings in line with the notion that a decrease in uncertainty regarding a patent's validity and scope facilitates the success of negotiations between patent sellers and potential buyers.

The following three chapters are self-contained and include their own introduction, conclusion, and appendix. Hence, each chapter can be read independently. References for all three chapters are listed in a joint bibliography at the end of this thesis.

Chapter 2

What to Buy when Forum Shopping – Determinants of Court Selection in Patent Litigation

2.1 Introduction

To ensure local accessibility to justice, most judicial systems are characterized by the coexistence of multiple geographically dispersed entry courts. If a dispute fulfills the requirements for territorial jurisdiction at more than one of these courts, the plaintiff gains the option to conduct *forum shopping*; that is, to freely select the court of his choice to seek judicial relief.[8] Particularly in patent litigation, forum shopping is considered common practice, since the infringing act, e.g., the manufacture or the sale of the infringing product, usually occurs at a national if not international scale (Moore, 2001b). The fact that plaintiffs actively select particular courts is initially surprising, as all courts in a judicial system are bound to the same substantive and procedural law. The legal realism literature therefore traces court differences back to the human factor in judicial decision making (cf. Posner, 1993; Stephenson, 2009). In particular, judges are assumed to differ in their expertise and ideology, leading to divergences from the legal norm that may favor one of the litigating parties. As a notable example, Atkinson *et al.* (2009) find evidence for the U.S. that plaintiffs deviate from their local district court if an alternative court seems less inclined to invalidate patents. The resulting disparity between

---

[8]Black's Law Dictionary defines forum shopping as "a litigant's practice of choosing the most favorable jurisdiction or court in which a claim might be heard." This definition emphasizes the litigation context, even though the notion of forum has been expanded to institutional certifiers, sponsors, and approvers (cf. Lerner and Tirole, 2006), but otherwise remains fairly broad. With help of the taxonomy by Bassett (2006), I define the kind of forum shopping dealt with in this study as domestic, horizontal forum shopping; that is, the deliberate selection of a particular court from a range of courts at the same instance with no differences regarding the applied substantive law and choice-of-law principles.

economic and litigious activity in some U.S. districts has triggered a steady stream of empirical legal literature that tries to link observable information on courts, such as average duration or win, settlement, and appeal rates, to their respective caseloads (see, e.g., McKelvie, 2007; Lemley, 2010; Lemley *et al.*, 2013; Lii, 2013). However, by relying on aggregated court level data, these studies disregard dispute-specific factors and their possible interactions with court characteristics, and fall short of fully explaining the popularity of particular venues among plaintiffs.

This study provides an initial analysis of the specific determinants of plaintiffs' court selection in patent litigation. Building upon the asymmetric information (*AI*) model of litigation and settlement by Bebchuk (1984), I depict court selection as an optimization problem for the plaintiff and derive predictions about his preferences. Even though theoretical economists are highly active in modeling disputes[9], the proceeding before court is mostly seen as failure in bargaining and denoted by an exogenous cost constant. Expanding the basic *AI* model, I implement court- and dispute-specific factors of the proceeding, such as opportunity costs due to delay in judgment and spatial distance, and analyze how these affect the plaintiff's expected utility of going to a particular court.

I test the propositions on the plaintiff's preferences using a comprehensive dataset on patent litigation in Germany. Providing fast and relatively cheap patent enforcement, German regional courts hear by far the largest share of patent litigation cases in Europe (Cremers *et al.*, 2013; Graham and van Zeebroeck, 2014).[10] The data cover all proceedings filed at three of the twelve available regional courts – namely, the regional courts in Düsseldorf (*DU*), Mannheim (*MA*), and Munich (*MU*) – between 2003 and 2008. These three courts account for approximately 80 to 90% of all patent infringement proceedings in Germany. Besides procedural details, the data also contain information on the identity of the litigants and the litigated patent at hand. In contrast to prior studies, I refrain from using aggregated case outcomes as indications of court heterogeneity, because a meaningful comparison of win, appeal, and settlement rates among courts is bound to the assumption of a random distribution of cases. This assumption can hardly be aligned with the premise of forum shopping, where the population of each court is conditional on the plaintiffs' self-selection. I therefore infer the main differences at each court exogenously from local procedural practices and the appointed judges. I derive the plaintiff's ex ante expected length of proceeding from each court by accounting for probabilities of two delaying events that may or may not occur during the proceeding. In particular, I predict the

---

[9]For surveys on the theoretical literature, see Spier (2007) and Daughety and Reinganum (2012).

[10]The German patent litigation system has gained global attention, particularly in the context of the ongoing *smartphone war* (cf. O'Brien, "German Courts at Epicenter of Global Patent Battles Among Tech Rivals," *New York Times*, 9 April 2012, 83).

alternative-specific likelihood of a requested expert opinion or a stay of the infringement proceeding due to a parallel validity challenge. This allows me to estimate the opportunity costs the plaintiff faces before legal enforcement of his patent. In addition, I capture the spatial distance between litigant and court as a further factor determining court selection. To recover estimates of the plaintiffs' preferences for the three courts, I use alternative-specific conditional logit models. While my focus is on the effect of variations in opportunity and transaction costs on the plaintiffs' preferences, I do not exclude court-specific differences in judicial decision making *prima facie*. The estimation models allow for court-specific biases that affect the expected utility of judgments.

In line with statements by practitioners (cf. Herr and Grunwald, 2011), I find court-specific likelihoods of delaying events. First, judges differ in their tendency to request an expert opinion depending on the patent's technology. The findings suggest that judges are specialized and have gained expertise in particular technologies from exposure to prior patent cases. I also identify differences in the judges' tendency to stay the proceeding due to a parallel validity challenge. Judges at the Düsseldorf regional court are significantly less likely to grant a stay than judges in Mannheim and Munich. Although ordinary proceedings take the longest at the Düsseldorf regional court due to its huge caseload, the low likelihood of delays often gives it the lowest expected length of proceeding.

The results of the alternative-specific conditional logit models support the theoretical predictions on the determinants of court selection. Opportunity costs due to delay in judgment have a significant negative effect on the plaintiff's court selection. This is particularly true if plaintiff and defendant are active in the same product market. Further, plaintiffs value local access to court. Transaction costs measured by distance to court have a significant negative effect on court selection. The magnitude of this effect is almost twice as large for small plaintiffs compared to large plaintiffs. I also find weak evidence that plaintiffs reach a general anti-patentee bias at the Munich regional court relative to the regional courts in Düsseldorf and Mannheim.

Even though forum shopping is considered a legitimate action authorized by law and repeatedly acknowledged by judges, scholars disagree on its welfare effects. Proponents link the welfare effect of forum shopping to the invisible hand argument. The plaintiff's free choice among multiple courts leads to efficiency gains in the market for litigation. That is, courts, facing institutional competition, have an incentive to invest in specialization, accrue experience and induce procedural innovations to attract fitting patent disputes, whereas plaintiffs are able to avoid congested dockets and courts lacking expertise in the patent's underlying technology (Moore, 2001b). Opponents of forum shopping, however, argue that forum shopping leads to systematic partiality in judicial decision making if one side in litigation, i.e., the plaintiff, dominates the decision on court selection. Legal scholars recently coined the term *forum selling* as the supply-side equivalent to forum shopping, referring to a court's leaning to decide pro-plaintiff to attract more cases (Klerman and Reilly, 2014).

Alternative designs for judicial systems that obstruct court competition mainly rely on either centralization, i.e., a single court hears all cases, or randomization, i.e., allocation of cases to courts is out of litigants' hands. These alternatives, however, likely disadvantage small litigants, who according to the results highly value local accessibility to court. This latter factor may be of particular importance for the discussion about the design of the Unified Patent Court (*UPC*), where the tradeoff between centralization versus local accessibility has been an abiding theme (Wadlow, 2015).

The study is structured as follows: Section 2.2 presents a model of the plaintiff's maximization problem, which provides predictions on the determinants of court selection. Section 2.3 describes the institutional framework of patent litigation in Germany. Section 2.4 provides details on the dataset and the construction of the variables. Section 2.5 contains the descriptive statistics. Section 2.6 then presents an econometric analysis of the determinants of court selection. Section 2.7 contains the conclusion.

2.2 A Model of Forum Choice

In this section I develop a model, built upon the litigation and settlement model by Bebchuk (1984), to depict court choice as an optimization problem for the plaintiff. Most theoretical analyses in the current literature on litigation and settlement are based on the premise of asymmetric information between plaintiff and defendant (Daughety and Reinganum, 2012). Bebchuk (1984) assumes one-sided asymmetric information between the litigants; that is, only the defendant knows the likelihood of being found liable by the court during the prejudgment bargaining phase.

I extend the basic model by introducing additional sources of court-specific costs of going to court, namely opportunity costs due to delay in judgment and the distance to court, affecting the plaintiff's expected utility of litigation. As a further feature of court heterogeneity, I assume the probability distribution of being found liable to differ among courts. From the comparative statics I then derive several predictions about the preferences of the plaintiff.

2.2.1 The Model

The expected utility of legal enforcement

Patent enforcement is a costly endeavor. I distinguish between three kinds of costs that arise through a patent infringement proceeding: opportunity costs, legal costs, and distance to court.

Opportunity costs: Opportunity costs emerge from the time required until the plaintiff is able to legally enforce his patent.[11] Every patent litigation dispute has a litigation value L that quantifies the economic scale of the infringement. This litigation value is determined by the value of the patent, the remaining patent protection period and the scope of infringement (cf. Section 2.3.2). I model L as the discounted value of all monopolistic rents the patent holder could hypothetically gain in the respective product market between the start of the dispute and the patent's expiration, T, if there was no infringement:

$$L(M,T,\delta) = \int_0^T M(1-\delta)^t dt$$

with M representing monopoly rents, discounted by the technology-specific factor $\delta (\delta \in [0,1])$.

The patented technology may be commercialized in multiple product markets.[12] If the plaintiff is active in a different product market, L will not reflect his actual means of appropriation in the infringer's product market. If the plaintiff is active in the product market of the infringing embodiment, he appropriates the patent through commercialization, which provides him with rent M. Alternatively, the plaintiff can license the patent to a firm active in that product market and receive rent B. I reasonably assume that $M > B$. I denote the product market proximity of the plaintiff to the infringer with the factor $\alpha (\alpha \in [0,1])$. At the extreme end of $\alpha = 1$, plaintiff and defendant are direct competitors in the same product market. In contrast, when $\alpha = 0$, the plaintiff is either active in a completely different product market or represents a non-producing entity (NPE).[13] Accordingly, the plaintiff's actual value from the patent in the infringer's market can be depicted as:

$$V(M,B,T,\delta) = \int_0^T (\alpha M + (1-\alpha)B)(1-\delta)^t dt.$$

---

[11]Opportunity costs in patent enforcement through legal means have gained little prior attention. Notable exceptions are Lanjouw and Lerner (2001) and Aoki and Hu (2003).

[12]For instance, the patented technology of a new aircraft braking system may also find application in the automobile industry.

[13]While I disregard the effect of third party competition, my reasoning follows in general the idea of 'rent effect' and 'dissipation effect' in the patent holder's decision to license out his patent (cf. Arora and Fosfuri, 2003). In particular, Gambardella and Giarratana (2013) show that licensing is more likely if the technology supports general purpose and the licensee is active in a different product market.

I assume the defendant continues her allegedly infringing action during the course of the patent litigation proceeding. Thus, in the time between filing the infringement action[14] and judgment, l, the plaintiff receives no rents at all. However, if the court considers the defendant liable for infringing the patent, the plaintiff can enforce his patent and enjoy rent $(\alpha M + (1 - \alpha)B)$ from then on. In addition, the plaintiff can demand compensation for the economic loss caused by the infringement. As will be outlined in Section 2.3.2, German courts grant damages that typically correspond to a calculatory license fee only, equal to B. The value of legally enforcing the patent W can then be expressed as:

$$W = \int_0^l (B)(1-\delta)^t dt + \int_l^T (\alpha M + (1-\alpha)B)(1-\delta)^t dt. \qquad (2.2.1)$$

Due to infringement and the inevitable delay in legal enforcement, the plaintiff therefore faces an irrecoverable loss of

$$C_o = V - W \qquad (2.2.2)$$
$$= \int_0^l (\alpha(M-B))(1-\delta)^t dt,$$

which I call the opportunity costs of patent enforcement through judgment. Following directly from above equation, opportunity costs increase with the length l of proceeding, the product market proximity between the litigants α, and the premium $M - B$.[15]

Legal costs: In Germany, legal costs, including the fees directly charged by the court and the litigants' legal fees, must be borne by the losing party. The court fees charged by the court are a function of the litigation value L; hence, these costs do not differ among courts.

Transaction costs: I further assume there are costs due to the proceeding that each litigant has to bear on their own, independent of the outcome of the proceeding. These costs may include, for instance, the costs of occupied human capital and the inconvenience of traveling to the court. I denote these as transaction costs C_{tp} for the plaintiff and C_{td} for the infringer. I assume these costs are a function of the spatial distance between the litigant and the selected court, R, as well as the size of the litigant, s. For $i = \{p, d\}$, I assume $\frac{\partial C_{ti}(R,s)}{\partial R} > 0$ and $\frac{\partial C_{ti}(R,s)}{\partial R^2} < 0$, as well as $\frac{\partial C_{ti}(R,s)}{\partial s} < 0$ and $\frac{\partial C_{ti}(R,s)}{\partial s^2} > 0$.

With the likelihood π of a judgment that allows the plaintiff to gain W from enforcing the

---

[14]For the sake of simplicity, I assume the plaintiff goes to court immediately after the infringement starts. I neglect the possibility of delayed discovery of the infringing act or lengthy pre-trial bargaining.

[15]There is little prior literature on the difference in patent rents between producing and non-producing firms, with the exception of Choi and Gerlach (2013).

patent, I express the plaintiff's expected utility from a judgment as

$$U_{Judgment} = \pi W - (1 - \pi) C_c - C_{tp}. \tag{2.2.3}$$

However, the plaintiff does not know the defendant's liability π, only its distribution. I denote the density function of the liability distribution with $f(\cdot)$ and the cumulative function with $F(\cdot)$. In line with Bebchuk (1984), I assume $f(\cdot)$ is positive within the interval $(a, b), 0 < a < b < 1$, and zero outside the interval. The function is continuous and differentiable throughout. I ignore cases in which the expected utility of judgment is negative, so the threat of filing an infringement action is always credible:

$$\pi W - (1 - \pi) C_c - C_{tp} > 0.$$

The option of settlement

Despite going to court, litigants frequently settle their patent dispute without a judgment. The litigation model by Bebchuk (1984) explicitly includes the option of settlement as an outcome of the bargaining process between the litigants. The bargaining process is depicted as a screening model in which the uninformed litigant, in my case the plaintiff, proposes a take-it-or-leave-it settlement offer S which the informed defendant has the option to either accept or reject. The defendant agrees to pay the settlement offer S if and only if it is not larger than her expected utility from a judgment that forces her to pay damages and an ongoing fee equal to the plaintiff's rents to pursue her activities,

$$S \le \pi (W + C_c) + C_{td}, \tag{2.2.4}$$

or, put differently:

$$\pi \ge \frac{S - C_{td}}{W + C_c}, \quad \text{with} \quad q(S) = \frac{S - C_{td}}{W + C_c} \quad \text{as the borderline case.}$$

The probability of the defendant accepting the settlement offer is therefore $1 - F[q(S)]$.

If the defendant is of a type lower than $q(S)$, the likelihood of her being held liable for infringement by the court is so small that she prefers to end the patent dispute through judgment. So, given the cases where she prefers to accept the settlement offer S, the conditional probability of the defendant being liable before court is $\frac{\int_a^{q(S)} x f(x) dx}{F[q(S)]}$.

The plaintiff's objective function of the patent dispute, which can either end in a settlement with the value S or a judgment with the expected utility of $U_{Judgment}$, therefore equals

$$U(S) = \underbrace{(1-F[q(S)])\,S}_{\text{Expected utility of settlement}} + \underbrace{F[q(S)]\left(-(C_{tp}+C_c)+(W+C_c)\frac{\int_a^{q(S)} xf(x)dx}{F[q(S)]}\right)}_{\text{Expected utility of judgment conditional on no settlement}}. \quad (2.2.5)$$

The plaintiff maximizes his utility of the patent dispute by choosing the optimal settlement offer, S^*. Following from the first order condition, the optimal settlement amount S^* must fulfill the equality:

$$1-F[q(S^*)] = \frac{C_{td}+C_{tp}+C_c}{W+C_c}\,f[q(S^*)]. \quad (2.2.6)$$

Accordingly, the settlement offer S^* is optimal if the marginal benefit due to the increase in S equals the marginal costs due to an increase in the likelihood of the need for a costly judgment. The second order condition is then:

$$f[q(S^*)] + \frac{C_{td}+C_{tp}+C_c}{W+C_c}\,f'[q(S^*)] > 0. \quad (2.2.7)$$

As the settlement offer S determines the settlement likelihood, or settlement rate, r, I denote the settlement rate corresponding to the optimal settlement offer S^* as

$$r^* = 1-F[q(S^*)]. \quad (2.2.8)$$

2.2.2 Comparative Statics

Aside from the settlement offer S, I have assumed every parameter of the plaintiff's objective function above to be exogenously given. I now propose that opportunity costs and spatial distance are court-specific, and hence the plaintiff selects the court that maximizes his utility. In addition to their direct effect on the expected value of a judgment, these court-specific cost variables can also affect the settlement rate and optimal settlement offer. Determining the net effect on the utility of the dispute is therefore less straightforward. The majority of my propositions follow reasoning by Bebchuk (1984).

I first assume heterogeneity among courts emerges due to different lengths of proceeding l. This causes the opportunity costs C_o to become court-specific if the activities of the litigants are not in perfectly different markets, i.e., $\alpha > 0$.

Proposition 1 *A decrease in opportunity costs, i.e., a higher W, always increases the plaintiff's expected utility of the dispute.*

This follows from equation (2.2.3), where I can see that the value of enforcement W increases with a decrease in opportunity costs C_o. The effect of the value of enforcement on the plaintiff's expected utility from the patent dispute is threefold. Following from equations (2.2.4) and (2.2.7), the expected utility of a judgment and the optimal settlement offer S^* increase with W. In contrast, the settlement rate r^* decreases with W as the borderline case $q(S^*)$ increases.

The plaintiff's own transaction costs C_{tp} become court-specific if the plaintiff's distance to one court is different from his distance to another.

Proposition 2 *A decrease in his own transaction costs C_{tp} always increases the plaintiff's expected utility of the dispute.*

This can be shown via the line of arguments used in Proposition 1. The expected utility of a judgment and the settlement S^* is affected negatively, but the settlement rate is affected positively. Thus, the overall effect is the same as from opportunity costs C_o.

Likewise, the defendant's transaction costs C_{td} may be court-specific if the defendant's distance to one court is different from her distance to another.

Proposition 3 *A decrease in the infringer's transaction costs C_{td} has an ambiguous effect on the plaintiff's expected utility of the dispute.*

The expected utility of a judgment remains unaffected by an increase in the defendant's transaction costs C_{td} (equation (2.2.2)), whereas the optimal settlement rate increases with C_{td} (equations (2.2.6) and (2.2.8)). The effect on the optimal settlement amount, however, is ambiguous. The infringer's transaction costs directly increase the optimal settlement amount S^*, but are also part of the function $q(S^*)$, which negatively affects the optimal settlement offer. Without a more specified liability distribution $F[\cdot]$, I cannot determine which effect is largest.

I finally assume that, along with costs, the liability distribution of π is also court-specific. I consider the defendant's liability to be dependent on the selected court. If, for instance, a court has a pro-patentee bias, the liability distribution undergoes an upward shift.

Proposition 4 *An increase in the defendant's liability π increases the plaintiff's expected utility of the dispute.*

This obviously increases the plaintiff's expected utility of a judgment (equation (2.2.4)) and the optimal settlement offer. Leaving the settlement rate constant, the upward shift unambiguously increases the plaintiff's expected utility of the dispute.

The results of the comparative statics are summarized in Table 2.1. As Proposition 3 is insufficient to provide clear predictions for the plaintiff's preferences, I will focus on Propositions 1, 2, and 4 in the empirical analysis.

Table 2.1: Comparative statics of model of forum choice

	Endogenous Variable			
Parameter	Expected utility of judgment	Optimal settle- ment amount	Likelihood of settlement	Overall effect on expected utility
Opportunity costs of plaintiff	−	−	+	−
Transaction costs of plaintiff	−	−	+	−
Transaction costs of defendant	0	?	+	?
Liability of defendant	+	+	0	+

2.3   Patent Litigation in Germany

2.3.1   Court Structure

Germany's court structure follows a continental style civil law system with a federal structure, which differentiates courts by both specialism and territory (Cabrillo and Fitzpatrick, 2008). Courts of general jurisdiction (*ordentliche Gerichte*) have authority over all civil disputes, including patent litigation, and constitute a four-tier hierarchy. The channel for patent litigation cases starts with the action filed at second-tier regional courts (*Landgerichte – LG*).[16] The number of regional courts with subject matter jurisdiction to hear patent litigation cases has been consolidated by the federal states, and currently stands at twelve (cf. Figure 2.4 in the Appendix).[17] To access any of these twelve regional courts, the plaintiff has to fulfill territorial jurisdiction requirements. The plaintiff can file his action at the jurisdiction of either the defendant's main place of business, residence, or the place of infringement. If the infringing act comprises a Germany-wide offer of the infringing embodiment[18], the plaintiff gains the op-

---

[16]Decisions by the regional courts can be appealed before their respective higher regional court (*Oberlandesgericht – OLG*; in Berlin: *Kammergericht*), and may be brought before the Patent Division of the Federal Court of Justice (*Bundesgerichtshof – BGH*) for a further appeal, limited to matters of law only.

[17]These are the regional courts in Berlin, Braunschweig, Düsseldorf, Erfurt, Frankfurt, Hamburg, Leipzig, Magdeburg, Mannheim, Munich, Nuremberg-Furth, and Saarbrücken. As the majority of these regional courts rarely see patent litigation cases, further consolidation has been proposed (Stauder, 1989). Interestingly, for competence reasons the former German Democratic Republic had all patent cases heard by a single court in Leipzig (Keukenschrijver, 1999).

[18]This includes, for instance, advertising on the internet or in national publications.

tion to file his action at any of the twelve regional courts. In reality, the plaintiff is usually unrestricted in his choice.[19]

Each of the regional courts has at least one chamber primarily designated to patent litigation cases.[20] A case is heard by a panel of three judges: one presiding and two sitting judges, all fully trained legal professionals. The plaintiff's claims must ground on a German patent (*DE*) or a European patent granted with effect for Germany (*EP*). The decision on infringement is enforceable throughout Germany.

2.3.2 The Infringement Proceeding

In contrast to patent litigation proceedings in other systems, proceedings on infringement before German regional courts are streamlined and have a clear, almost rigid outline that allows little divergence from the ordinary structure.[21] Judges usually refrain from stepped actions and instead initiate separate, adjacent proceedings for additional claims (cf. Figure 2.5 in the Appendix). In the following, I will focus on the heart of a patent dispute – the infringement main proceeding.

Structure of the main proceeding

The infringement's main proceeding is initiated by the plaintiff through filing the infringement action, in which he states his claims and estimates the litigation value (cf. Figure 2.6 in the Appendix). Several forms of legal relief are available to the plaintiff. He may claim for an order to cease and desist from further infringement, for recall and destruction of the infringing goods, for information and rendering of account to identify distribution channels and calculate damages, for compensation of damages, as well as for notification of judgment (Kühnen, 2012, pp. 266 et seqq.). Subject to the court's practice, the litigants meet in a so-called early oral hearing, where deadlines for the further exchange of statements and the date for the main oral hearing are scheduled. Alternatively, the court gives notice in written form. Subsequently, the alleged infringer states her defense. In contrast to other systems, the alleged infringer cannot challenge the patent's validity in the infringement proceeding (cf. Cremers *et al.*, 2014).

---

[19]Unlike in the U.S., the defendant has no legal means to demand a transfer if the current court's requirements for territorial jurisdiction are met. Furthermore, the prior request for a declaratory judgment by the alleged infringer does not restrict the patent holder from subsequently filing his action at another suitable regional court. Rather, the request for declaratory judgment will be terminated and become part of the proceeding initiated by the patent holder.

[20]If the regional court has more than one chamber designated to hearing patent cases, the internal assignment of the filed action follows a transparent system specified in the court's case assignment plan. However, this system is unpredictable ex ante for the plaintiff.

[21]See Harguth and Carlson (2011) or Kühnen (2012) for an elaborate description of the German patent infringement proceeding.

Prior to the main oral hearing, each party exchanges between one and two written statements specifying their own reasoning and countering the opposing party's arguments. The main oral hearing takes place roughly between six to twelve months after the action was filed, primarily depending on the court's docket. The judges give written notice to the litigants about their judgment usually four to eight weeks after the main oral hearing, concluding the proceeding in the first instance.

Infringement proceedings mainly diverge from the structure described above, if the judges decide, during or after the main oral hearing, to either stay the proceeding due to a parallel invalidity proceeding or demand further evidence in the form of an expert opinion. Both events will considerably delay the judgment on infringement.

Expert opinion

Construing patent claims and analyzing the composition of the allegedly infringing embodiment requires a sophisticated understanding of the respective technology from the presiding judge (Kühnen, 2012, p. 562). Judges experienced in dealing with patent infringement cases can answer technical questions independently if they have the necessary expertise. However, if the facts are technically complex and the judges lack the technical expertise to decide on infringement, they must request an expert opinion. The decision to request an expert opinion is at the judges' discretion; however, judges are advised to rely on experts if their own expertise is insufficient.[22] The litigants have very limited influence on the request for an expert opinion. The call for an expert by the judge can be neither ordered nor challenged by the litigants. The experts appointed to state their opinion for the assessment of technical questions in written form are usually professors or patent attorneys with a significant expertise in the respective field of technology. The request for an expert opinion usually delays a decision on infringement by up to two years (Kühnen, 2012, p. 562).

Stay of proceeding

The alleged infringer may request to stay the infringement proceeding until a decision concerning a parallel patent validity challenge becomes available. The effect of a validity decision on the outcome of an infringement proceeding can be significant. If the validity challenge is entirely successful, the patent will be declared *ex tunc* invalid and any pending infringement proceeding will be discontinued. If the patent is partly revoked, the subject matter in the infringement proceeding will be considered based on the amended patent. A parallel invalid-

---

[22]In fact, it may constitute grounds for an appeal to the Federal Court of Justice if a judge clearly overestimates his understanding of certain aspects of the case (Kühnen, 2012, p. 566).

ity proceeding can arise due to an opposition filed before the European Patent Office for *EP* patents, or before the German Trade Mark and Patent Office for *DE* patents. After the opposition phase, invalidity proceedings for both kinds of patents are initiated through a revocation action filed before the German Federal Patent Court.

The delay to a judgment due to a stayed proceeding can be considerable. The German Federal Patent Court decides on validity in sixteen to twenty-four months. Including appeal, litigants have to expect a maximal length of up to five years until a final judgment on validity is given (Cremers *et al.*, 2014). Likewise, oppositions may take between three to four years.

Damages

The plaintiff may demand compensation for economic loss due to the infringement. The question of the level of damages is usually not part of the main proceeding, but answered in a separate, subsequent proceeding. Three methods of calculating damages are available: based on the plaintiff's lost profits, on the infringer's gained profits, or per license analogy. The plaintiff is free to choose the method, independent of his status or market activities (Kühnen, 2012, p. 527). The calculation method based on the plaintiff's lost profits is rarely applied in proceedings on the amount of damages. This is mainly due to the plaintiff's requirement to disclose his accounts in the proceeding and the challenge to provide evidence for causality between the infringement and unrealized profits. Likewise, plaintiffs consider compensation based on the infringer's profit an unpopular choice, as the infringer is able to minimize her profits through the inclusion of overhead costs. Accordingly, the license analogy calculation is the most widely used (Schramm and Kaess, 2010, p. 377; Kühnen, 2012, p. 547). Here, the amount of damages is calculated based on what the infringer would have had to pay as reasonable fees if she had entered a license agreement. Calculation based on license analogy is considered a simple, convenient method, but usually constitutes the lower limit of the plaintiff's economic loss due to infringement (Müller-Stoy and Schachl, 2011, p. 342).

In comparison with pro-patentee damages rules applied in other countries, such as the U.S. or France (cf. Love, 2009; Cotter, 2013), compensation claims remain a barely effective part of patent enforcement in Germany.

2.4   Data and Construction of Variables

2.4.1   Data

To empirically test my predictions regarding the determinants of court selection, I draw upon a dataset of patent litigation proceedings filed between 2003 and 2008 in Germany. I use several additional data sources to complement the dataset with respect to the characteristics of the litigants, the courts, and the patents in dispute.

Infringement proceeding

I collected the data on infringement proceedings directly from court records stored at the three regional courts covering the most patent litigation cases in Germany: the Düsseldorf, Mannheim, and Munich regional courts.[23] The dataset covers information on procedural aspects, the identity of the litigants and their legal representatives, and the patents at issue. In particular, I obtained information about when and how the proceedings ended; that is, by judgment, settlement, or withdrawal. If there was a judgment, I learned the outcome (win, partial win, loss) and whether an appeal was filed. I also acquired information on the litigation value set by the court and the claims made by the plaintiff, which helped me to identify noninfringement or adjacent proceedings (cf. Section 2.3.2).

Litigants

The data also include names and addresses of the litigants and their legal representatives. After matching the names of corporate litigants to firm level databases, including Bureau van Dijk's Orbis, Compustat, and THOMSON One, I complemented the data with information on the litigants' fundamentals (number of employees, total assets, and turnover) and industry activities (NACE Rev. 2 industry codes).[24] The data allow to distinguish between natural and legal entities, such as firms, research institutions, universities, etc. I also identified non-producing entities (NPEs) among corporate litigants in accordance with the methodology introduced in Helmers *et al.* (2014).

Patents

I unambiguously identified nearly all litigated patents, since they are referenced in the case records by their application (and publication) numbers. Using PATSTAT, I retrieved biblio-

---

[23]For details on the collection process, see Cremers *et al.* (2013).
[24]For firms without an entry in any firm level database, I manually added information from online sources.

graphic and procedural information on the patents, such as application and examination dates, IPC classifications, equivalents, and patent as well as nonpatent references. In addition, legal status information from PATSTAT helped me to identify post-grant oppositions against the litigated patents. In particular, I have acquired information on the filing and ruling dates and the binding outcome of the opposition. Data on revocation proceedings for the litigated patents had to be collected separately. I extracted information from judgments by the Federal Patent Court and its appeal court, the Federal Court of Justice.[25] I also gained information on the filing dates and withdrawals of revocation actions in both instances from the the German Patent and Trademark Office register. This allowed me to reconstruct the course of the revocation proceedings even without access to the respective court records.

Court

I complemented the dataset with further information about the three regional courts, based on the regional courts' annual case assignment plans. The case assignment plan defines the subject matter each chamber will hear and how cases are allocated if more than one chamber can hear the case. The case assignment plan also designates each chamber's presiding judge and the pool of sitting judges. I obtained biographical data on the presiding judges, i.e., age, current and prior positions, and courts of employment. This information is publicly available via the various editions of the handbook on Germany's judicial system (*Handbuch der Justiz*), which is published biennially by the German association of judges.

2.4.2 Construction of Variables

In the following, I briefly describe the variables constructed from the dataset. I distinguish between variables which capture characteristics of the patent, the court, and the dispute.

Patent characteristics

As discussed in Section 2.3.2, infringement proceedings may be subject to delay due to a stay of proceedings or the request for an expert opinion. The request for an expert opinion largely depends on the intricacy[26] of the litigated patent, while the grant of a stay of proceedings depends primarily on the legal quality[27] of the litigated patent. Earlier literature uses indicators constructed from patent information primarily as determinants of patent value. Although I am

---

[25]Both courts publish all their decisions on validity since 2000 on their websites.

[26]The technological *complexity* of a patent commonly refers to the cumulative nature of the invention (Cohen *et al.*, 2000). I therefore use the term *intricacy* to avoid confusion.

[27]I disregard the techno-economic aspects of the patent's underlying invention and focus on the legal quality of the patent's certainty in terms of scope and enforceability (cf. Thomas, 2002; Burke and Reitzig, 2007).

aware of the potential lack of discriminatory power, I also capture two other patent characteristics: intricacy and patent (cf. Table 2.2 for an overview).

Table 2.2: Indicators for intricacy, quality and value of patents

Patent indicators	Intricacy	Quality	Value	Basic references
Bibliographical				
Backward citations (patents)			✓	Harhoff *et al.* (2003)
Backward citations (nonpatent literature)	✓		✓	Carpenter *et al.* (1981); Harhoff *et al.* (2003); Cassiman *et al.* (2008)
Forward citations		✓	✓	Trajtenberg (1990); Harhoff *et al.* (1999); Lanjouw and Lerner (2001)
No. of claims	✓		✓	Lanjouw and Schankerman (2001); Harhoff *et al.* (2003); Harhoff and Wagner (2009)
IPC count	✓		✓	Lerner (1994); Harhoff *et al.* (2003); Harhoff and Wagner (2009)
Family size		✓	✓	Lanjouw *et al.* (1998); Harhoff *et al.* (2003)
PCT filing			✓	Guellec and van Pottelsberghe de la Potterie (2000)
EP bundle patent		✓	✓	Guellec and van Pottelsberghe de la Potterie (2000)
Age of patent (since filing)	✓	✓	✓	Lanjouw *et al.* (1998); Lanjouw and Lerner (2001)
Procedural (pre-grant)				
Accelerated examination requested			✓	Harhoff and Reitzig (2004); Harhoff and Wagner (2009)
Grant lag	✓		✓	Harhoff and Wagner (2009); Régibeau and Rockett (2010)
Procedural (post-grant)				
Patent solidified through opposition proc.		✓	✓	Harhoff *et al.* (2003); Cremers *et al.* (2014)
Patent challenged through revocation proc.		✓	✓	Cremers *et al.* (2014)
Patent solidified through revocation proc.		✓	✓	Harhoff *et al.* (2003); Cremers *et al.* (2014)

Intricacy: The intricacy of a patent is primarily derived from two sources. The first is the
depth and specificity of the patented technology. The second is the degree of originality and
detachment of the patented technology from established technologies. Both characteristics
make it hard for laypersons, such as judges, to comprehend the patent at issue and define its
scope. For judges lacking the appropriate academic background, understanding the technical
aspects of a patent based on science rather than established technologies is more demanding.
Accordingly, I capture the intricacy of a patent via the ratio between nonpatent literature citations, which consist primarily of scientific literature (Callaert *et al.*, 2006), and total citations.

As granting of a patent requires novelty and an inventive step compared to the current state of art, technologies underlying recently granted patents are less likely to have already entered the domain of common knowledge. I therefore include the age of the patent as a further measure of intricacy. Likewise, Régibeau and Rockett (2010) argue that grant lag is largest for patents at an early stage of their innovation cycles and decreases as technologies mature. I include the grant lag normalized to the average length of examination at the respective patent office as a further measure of intricacy. Aside from technology, the intricacy of a patent may also derive from the breadth of patent scope. The breadth of a patent is commonly operationalized as the number of claims and the number of assigned IPC subclasses (Novelli, 2015). Harhoff and Reitzig (2004) argue that the number of claims reflects the scope, and thus, intricacy of a patent. Likewise, Lerner (1994) draws on the number of IPC subclasses to capture patent scope. I therefore include these two measures as well.

Quality: A patent's quality is best measured by how well it fulfills the statutory requirements of patentable subject matter, novelty, nonobviousness and disclosure. Unfortunately, I am unable to analyze these criteria to determine the likelihood of invalidity and must therefore rely on the outcomes of prior invalidity proceedings. In line with Cremers *et al.* (2014), I assume that patents that survived an opposition or revocation proceeding have solidified their validity. In fact, infringement courts base their assessment of the likelihood of validity on the outcomes of prior invalidity proceedings and rarely stay a proceeding if the prior art used in the validity challenge has been referenced in prior invalidity proceedings (Kaess, 2009; Scellato *et al.*, 2011). I also distinguish patents by application authority, since scrutiny in examining patents can differ between the EPO and DPMA. I further include forward citations and the age of the patent as measures for legal quality, because both variables capture the social diffusion of the patent (Lanjouw and Lerner, 2001) and potential validity challenges.

Value: According to Cremers *et al.* (2014), the value of a patent correlates with the likelihood of a validity challenge, but is not supposed to be part of a judge's consideration to stay the proceeding or request an expert opinion. To capture value, I rely on established indicators such as backward citations to patents (Harhoff *et al.*, 2003) and nonpatent literature, forward citations, and the number of assigned IPC subclasses (Harhoff *et al.*, 2003). The value of a patent is also reflected in the costs the patent holder is willing to bear to gain and maintain protection. The costs of a patent are determined by the number of claims, the decision to file the patent via the PCT route and apply for an *EP* bundle patent (Lanjouw *et al.*, 1998; Guellec and van Pottelsberghe de la Potterie, 2000). In line with this, the geographical (family size) and temporal (age) scope of protection serve as additional measures of patent value (Lanjouw *et al.*, 1998).

Technology area: Patent intricacy, quality and value likely differ across technology areas. I therefore map the IPC codes assigned to the patents in line with the concordance table developed by the *Fraunhofer ISI* and the *Observatoire des Sciences et des Technologies* in cooperation with the French patent office (cf. Schmoch, 2008). The IPC codes are clustered into five primary technology areas: (a) electrical engineering, (b) instruments, (c) chemistry, (d) mechanical engineering, (e) other fields.

Court characteristics

Expertise of judges: According to Ann (2009), German judges are mostly self-educated in technical matters. Moore (2001a) argues that judges gain technical expertise primarily from frequent exposure to the technology. I follow the general approach of Kesan and Ball (2011), who measure expertise by the prior caseload and seniority of a judge. Formally speaking, I define the expertise of a judge g in a certain technology x in year y of his tenure as the judge's prior exposure to that technology area since the beginning of his presidency. I operationalize this exposure as the sum of all prior patent infringement disputes I with a patent of technology x that required the judge's involvement:

$$Prior\ exposure_{gxy} = \log\left(\sum_{y=0}^{Y-1}\sum_{i=1}^{I} Case_{gixy} + 1\right).$$

Since judges may also benefit from their general experience in patent litigation, I include the tenure of the presiding judge in years as an additional variable. I also take into account whether the judge can draw on prior infringement decisions based on the same patent.

Dispute characteristics

In his court selection problem, the plaintiff has to consider the opportunity costs he faces at each court. These costs are a function of the time the alleged infringer is able to continue her activities, the product market proximity, and the litigation value.

Expected length of proceeding: The length of proceeding ex ante expected by the plaintiff is a latent variable for several reasons. On account of this, I predict the expected length for each proceeding at all three courts (cf. Section 2.6.1 for methodological details).

Product market proximity: The plaintiff's opportunity costs also depend on the litigants' product market proximity to each other.[28] I use the overlap of the corporate litigants' market

---

[28]In the patent literature, competition between two parties is commonly measured as the overlap of their patent portfolios or a derivative of it. I refrain from using this method, because it is not applicable to litigants with no patents on their own and captures technological rather than product market proximity.

activities captured by industry codes available from firm level databases. The constructed product market proximity variable is discretized with distinct values between 0 and 1, with 1 reflecting perfect overlap of market activities. Natural persons as well as nonpracticing entities, including research institutions, universities and patent assertion entities, are by definition not active in any product market. Therefore, their product market proximity is always 0.[29]

Litigation value and legal costs: The litigation value as set by the court is referenced in the case files. In the few cases where the litigation value was adjusted during the proceeding, I choose the most recent one. Although the data partly include details on the exact legal costs, I determine legal costs for all cases as a function of the litigation value stated in the court records. With the help of a legal costs calculator,[30] I calculate the approximate costs for any litigation value of the proceedings.

Transaction costs: I operationalize transaction costs as the spatial distance between the litigant's (business) address and the court's address. In case of multiple plaintiffs or defendants, I choose the one with the shortest distance to the court. I calculate the variable by retrieving longitudinal and latitudinal data through the Google Maps API (Ozimek and Miles, 2011). For litigants from outside of Continental Europe, travel distance is not calculable. Here, I use an approximate distance based on the flight distance between the courts and each of the litigants' countries.

Control variables

Size and residence of litigants: To account for the litigant's stress due to the occupation of financial and human resources in the proceeding, I include a variable capturing the size of each litigant. The size categories follow the EU definition and rest upon the litigant's number of employees, turnover, and total assets. I assume natural persons as having one employee. Fundamentals of research institutions and universities were manually researched and added to the data. In case of multiple plaintiffs or defendants, I give the largest party priority.[31] I take the residence of the litigants into account and distinguish between German, European and non-European litigants.

Legal representatives: While court selection is an essential part of pretrial strategy (Stieger, 2004), legal representatives likely have different information on court-specific characteristics available to them. In particular, small law firms or self-employed attorneys probably

---

[29]Further details on how I derive product market proximity from the industry codes can be found in Section 2.8.3 in the Appendix.

[30]The legal costs calculator I used in this study can be accessed online: http://foris-prozessfinanzierung.de/Prozesskostenrechner [accessed: 22 July 2015].

[31]As an alternative measure, I gathered the fundamentals for all parties on one side. The results did not change significantly.

lack the resources and knowledge to identify the optimal court. I include a dummy variable indicating whether the plaintiff or defendant's legal representative is considered a top law firm for patent litigation in Germany. I use the annual ranking published by the professional journal *JUVE Rechtsmarkt* in 2009.[32]

Multijurisdictional litigation: The rationale for court selection in patent disputes where the litigants have encountered each other before court in multiple countries is very likely to differ. I therefore identify multijurisdictional litigation by matching patent numbers and litigants from my data with available litigation data for the UK (England and Wales), France, and the Netherlands (cf. Cremers *et al.*, 2013).

2.5   Descriptive Analysis

Sample description

The data from the Düsseldorf, Mannheim and Munich regional courts during the period 2003 to 2008 contain 4,060 proceedings. Most of these cases were filed before the Düsseldorf regional court (2,534 cases). The Mannheim regional court is next with 1,196 and followed by the Munich regional court with 330 cases. With few exceptions, these cases are all patent litigation cases in the broadest definition.

For the purpose of this study, I want to achieve a homogeneous sample of patent infringement main proceedings with complete information on all essential procedural features (cf. Table 2.3). I therefore identified and removed cases in which the subject matter suggests that the litigants were previously in a contractual relationship. This includes cases on employee invention disputes, licensing, assignment, and patent transfer disputes.[33] I also dropped infringement actions based on utility models. As my focus is on patent infringement main proceedings, I removed single preliminary injunctions and adjacent proceedings, such as cost or damages proceedings (cf. Section 2.3.2). The resulting sample consists of patent infringement main proceedings, which I merged with additional information on patent characteristics and litigants. I then identified several cases where the patent was either not yet granted or not in force at the time the infringement action was filed. Excluding those left 2,599 infringement proceedings. Eventually, I regress the court selection only on cases without a prior preliminary injunction. Considerations for forum shopping in preliminary injunction requests likely differ from those in patent infringement proceedings. Moreover, the infringing act can be halted

---

[32]As an alternative measure, I classified top legal representatives as those law firms that represent more cases than the average law firm in the data. The two measures were highly correlated.

[33]I assume that with any prior contractual relationship, the main issue is unlikely to be the patent itself.

through the granting of a preliminary injunction, to the effect that the plaintiff is relieved of most opportunity costs.

Table 2.3: Overview and definition of subsamples

Sample definition	Regional court			Total
	LG DU	LG MA	LG MU	
All proceedings				
N	2,534	1,196	330	4,060
% of subsample / full sample	65.38	26.78	7.83	100.00
– with patent infringement				
N	1,773	695	198	2,666
% of subsample / full sample	66.50	26.07	7.43	65.67
– based on patents in force				
N	1,719	692	188	2,599
% of subsample / full sample	66.14	26.63	7.23	64.01
– without prior preliminary injunction				
N	1,353	621	137	2,111
% of subsample / full sample	64.09	29.42	6.49	52.00

Notes: The unit of observation is the (infringement) proceeding. Patents in force are defined as patents that are granted, yet neither expired nor bindingly invalidated at the point of filing.

Descriptive statistics

In the following I present aggregated court data. First, Figure 2.1 illustrates the distribution of proceedings filed between 2003 and 2008 across the three regional courts. Düsseldorf comfortably leads the field with Mannheim and Munich as the second and third busiest courts, respectively. Neither the distribution nor the settlement rate shows heavy fluctuation for the time frame. Notably, the caseload in Mannheim increases significantly after the establishment of the second chamber in 2005.[34]

Comparing the outcomes of proceedings across court, Table 2.4 implies that the Düsseldorf regional court rules that patents have been infringed more often than Mannheim. Taking into account the varying settlement rates, Mannheim has the highest win-rate for patent holders.

Looking at the densities of the length of proceeding by court in Figure 2.2, I observe that the three courts significantly differ in the time needed until judgment. The length of ordinary

---

[34]A causal relationship between the establishment of the second chamber in Mannheim and the drop in cases at the Düsseldorf regional court in 2006 is in line with statements made by interviewed practitioners.

Figure 2.1: Number of proceedings with judgment and settlement by court and year

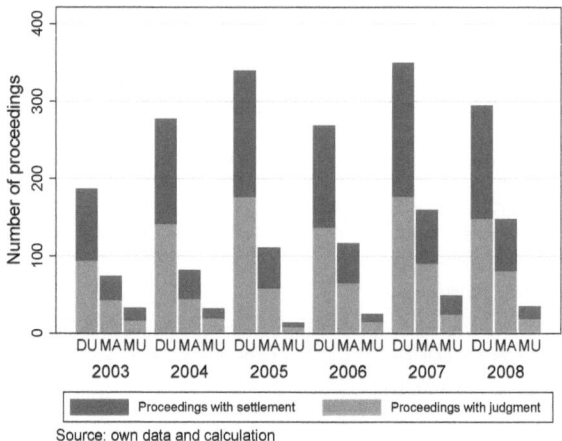

Source: own data and calculation

Notes: The sample consists of all patent infringement proceedings based on patents in force. The unit of observation is the infringement proceeding. Judgments include judgments by contention and by default decree. Settlements include withdrawals as well as (out-of-court) settlements.

Table 2.4: Outcomes of infringement main proceedings by regional court

	Regional court							
	LG DU		LG MA		LG MU		Total	
Outcome	N	%	N	%	N	%	N	%
infringed	487	28.6%	142	20.6%	49	26.5%	678	26.3%
partly infringed	146	8.6%	26	3.8%	17	9.2%	189	7.3%
not infringed	273	16.0%	66	9.6%	33	17.8%	372	14.4%
settlement	795	46.7%	455	66.0%	86	46.5%	1,336	51.9%
Total	1,701	100.0%	689	100.0%	185	100.0%	2,575	100.0%

Notes: The sample consists of all patent infringement proceedings based on patents in force. The unit of observation is the infringement proceeding. 24 observations without information on outcome. Judgments include judgments by contention and by default decree. Settlements include withdrawals as well as (out-of-court) settlements. The outcome is defined from the plaintiff's perspective.

proceedings can be easily identified for the Düsseldorf and Mannheim regional courts. The high density of 400 days indicates the average length of an ordinary main proceeding with judgment in Düsseldorf. Ordinary main proceedings in Mannheim end on average after 280 days. The relatively fat tail to the right for proceedings in Mannheim, however, indicates a higher likelihood of delaying events compared to Düsseldorf. At all three courts, a considerable share of settlements occurs in the first 100 days. Despite having early oral hearings, most settlements at Düsseldorf occur later than at Mannheim and Munich. As I observe frequent settlements after the point when ordinary proceedings usually end, I conclude that not all proceedings with a granted stay or a requested expert opinion eventually result in a judgment.[35]

Figure 2.2: Length of infringement main proceedings with judgment and settlement by court (densities)

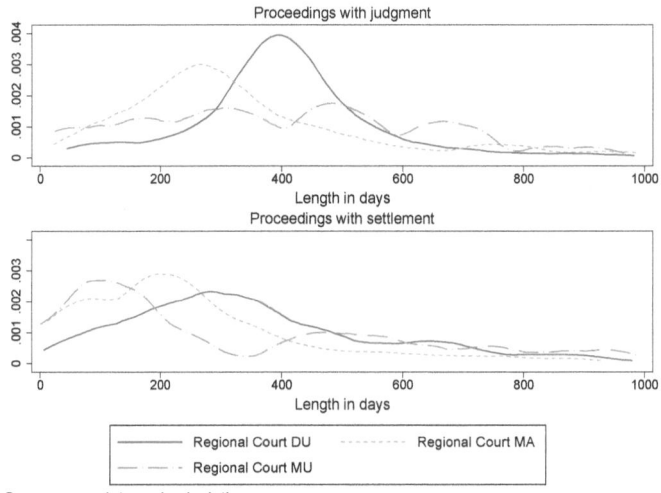

Source: own data and calculation

Notes: The sample consists of all patent infringement proceedings based on patents in force. The unit of observation is the infringement proceeding. Judgments include judgments by contention and by default decree. Settlements include withdrawals as well as (out-of-court) settlements. Truncated at 1,000 days.

Overall, I am not able to explain why the regional court in Düsseldorf attracts considerably more plaintiffs than the regional courts in Mannheim and Munich using court level data alone.

I now turn to a brief description of the summary statistics of the main variables in Table 2.5. Several significant differences are observable across the regional courts. Most notably, the litigation value is over twice as high in Düsseldorf on average than at the other two courts. This difference persists across all technology main areas (cf. Table 2.11 in the Appendix). I observe a parallel invalidity proceeding in about 50% of all proceedings at the Düsseldorf regional court. The rate is even higher in Mannheim, at 60%, but under 40% in Munich. Conditional on such a parallel invalidity proceeding, the rate of a granted stay is lower for Düsseldorf and Mannheim, at 18% compared to 30% in Munich. Likewise, expert opinions are requested about twice as often in Munich (20%) as in Düsseldorf (8%) or Mannheim (12%). Multijurisdictional litigation, where the litigants also face each other before court in another

---

[35] I broke down the lengths of proceedings with settlements and judgments for each year and found fluctuations but no clear time trend among the courts (cf. Figure 2.7 and Figure 2.8 in the Appendix).

European jurisdiction, is a relatively rare event and shows little difference in frequency among the courts.

Table 2.5: Summary statistics grouped by regional court

	Regional court								
	LG DU			LG MA			LG MU		
	Mean	Min	Max	Mean	Min	Max	Mean	Min	Max
Courts									
Judgment (at first instance) (d)	0.54	0	1	0.34	0	1	0.54	0	1
— plaintiff wins (d)	0.54	0	1	0.61	0	1	0.49	0	1
— plaintiff partly wins (d)	0.16	0	1	0.11	0	1	0.17	0	1
— plaintiff loses (d)	0.30	0	1	0.28	0	1	0.33	0	1
— appealed (d)	0.39	0	1	0.30	0	1	0.34	0	1
Judgment by consent/default decree (d)	0.12	0	1	0.36	0	1	0.24	0	1
Litigation value (in thousand €)	1,093.41	1	30,000	419.86	0	16,500	333.64	5	3,500
Court fees (in thousand €)	13.52	0	274	6.36	0	153	5.63	0	36
Total legal costs (in thousand €)	42.42	1	928	21.45	0	456	19.34	2	107
Length of proceeding (in months)	13.18	0	123	10.70	0	88	12.20	0	78
Parallel opposition proceeding (d)	0.17	0	1	0.12	0	1	0.09	0	1
Parallel revocation proceeding (d)	0.32	0	1	0.49	0	1	0.28	0	1
— infringement proceeding stayed (d)	0.18	0	1	0.18	0	1	0.30	0	1
Expert opinion (d)	0.08	0	1	0.12	0	1	0.20	0	1
Preliminary injunction (d)	0.21	0	1	0.10	0	1	0.27	0	1
Multijurisdictional litigation (d)	0.03	0	1	0.01	0	1	0.02	0	1
Judges									
Tenure as judge (in years)	11.80	7	16	15.68	2	25	22.05	15	30
Prior exposure to technology area	5.10	3	6	4.04	1	5	2.77	0	4
Judge 1 (LG DU) (d)	0.50	0	1	0.00	0	0	0.00	0	0
Judge 1 (LG MU) (d)	0.00	0	0	0.00	0	0	0.44	0	1
Judge 1 (LG MA) (d)	0.00	0	0	0.21	0	1	0.00	0	0
Judge 2 (LG DU) (d)	0.41	0	1	0.00	0	0	0.00	0	0
Judge 2 (LG MU) (d)	0.00	0	0	0.00	0	0	0.56	0	1
Judge 2 (LG MA) (d)	0.00	0	0	0.43	0	1	0.00	0	0
Judge 3 (LG MA) (d)	0.00	0	0	0.36	0	1	0.00	0	0
Judge 3 (LG DU) (d)	0.09	0	1	0.00	0	0	0.00	0	0
Patent holder									
Nonpracticing entity (d)	0.11	0	1	0.46	0	1	0.27	0	1
Micro (d)	0.09	0	1	0.13	0	1	0.24	0	1
Small (d)	0.10	0	1	0.08	0	1	0.11	0	1
Medium (d)	0.16	0	1	0.14	0	1	0.18	0	1
Large (d)	0.65	0	1	0.65	0	1	0.47	0	1
Germany (d)	0.47	0	1	0.40	0	1	0.62	0	1
Europe (excl. Germany) (d)	0.28	0	1	0.53	0	1	0.29	0	1
World (excl. Europe) (d)	0.25	0	1	0.07	0	1	0.09	0	1
Distance to court (in thousand km)	2.13	0	17	0.92	0	9	1.13	0	9
Top legal representative (no. of cases) (d)	0.89	0	1	0.79	0	1	0.61	0	1
Top legal representative (JUVE) (d)	0.84	0	1	0.27	0	1	0.42	0	1
Alleged infringer									
Micro (d)	0.09	0	1	0.24	0	1	0.18	0	1
Small (d)	0.15	0	1	0.17	0	1	0.27	0	1
Medium (d)	0.25	0	1	0.25	0	1	0.24	0	1

Continued on next page

Table 2.5 – continued from previous page

	LG DU			LG MA			LG MU		
Large (d)	0.51	0	1	0.33	0	1	0.31	0	1
Germany (d)	0.72	0	1	0.81	0	1	0.66	0	1
Europe (excl. Germany) (d)	0.19	0	1	0.09	0	1	0.19	0	1
World (excl. Europe) (d)	0.08	0	1	0.10	0	1	0.14	0	1
Distance to court (in thousand km)	0.94	0	17	1.12	0	15	1.65	0	12
Top legal representative (no. of cases) (d)	0.78	0	1	0.63	0	1	0.70	0	1
Top legal representative (JUVE) (d)	0.51	0	1	0.28	0	1	0.33	0	1
Product-market proximity	0.70	0	1	0.47	0	1	0.60	0	1
Patent characteristics									
No. of patents in proceeding	1.10	1	9	1.06	1	7	1.15	1	4
EP bundle patent (d)	0.81	0	1	0.79	0	1	0.75	0	1
PCT filing (d)	0.28	0	1	0.16	0	1	0.25	0	1
Forward citations (in first 5 years)	3.14	0	51	6.66	0	41	3.22	0	15
Backward citations (patents)	5.12	0	32	5.78	0	33	5.12	1	27
Backward citations (nonpatent literature)	0.88	0	41	1.94	0	21	1.03	0	21
Nonpatent literature ratio	0.11	0	1	0.18	0	1	0.10	0	1
No. of claims	14.34	1	158	16.30	1	48	14.01	1	75
IPC subclass count	2.05	1	11	4.31	1	9	2.69	1	9
Family size (INPADOC)	10.96	1	183	28.54	1	69	14.31	1	69
Year of application/priority	1993.75	1980	2005	1993.01	1983	2004	1994.61	1981	2004
Year of patent grant	1998.21	1982	2008	1997.28	1986	2007	1998.26	1985	2007
Grant lag (difference from mean in days)	99.40	-1,303	4,338	40.33	-1,162	3,972	-136.49	-1,045	2,411
Accelerated examination requested (d)	0.19	0	1	0.12	0	1	0.21	0	1
Age of patent (in years)	11.86	2	25	12.86	2	23	11.04	2	22
Prior infringement decision on patent (d)	0.10	0	1	0.07	0	1	0.12	0	1
Patent technology area									
Chemistry (d)	0.18	0	1	0.07	0	1	0.10	0	1
Electrical engineering (d)	0.25	0	1	0.49	0	1	0.24	0	1
Instruments (d)	0.12	0	1	0.08	0	1	0.07	0	1
Mechanical engineering (d)	0.30	0	1	0.21	0	1	0.30	0	1
Other (d)	0.16	0	1	0.14	0	1	0.29	0	1
Patent invalidation history									
Patent solidified (opposition proc.) (d)	0.15	0	1	0.09	0	1	0.16	0	1
Patent challenged (revocation proc.) (d)	0.08	0	1	0.34	0	1	0.16	0	1
Patent solidified (revocation proc.) (d)	0.02	0	1	0.02	0	1	0.03	0	1
N	1,719			692			188		

Notes: The sample consists of all patent infringement proceedings based on patents in force. The unit of observation is the infringement proceeding. Judgments include judgments by contention and by default decree. Settlements include withdrawals as well as (out-of-court) settlements.

While the average size of patent holders is significantly smaller at the Munich regional court compared to the other two courts, the share of large alleged infringers is greatest at the Düsseldorf regional court. Notably, the Mannheim regional court attracts by far the largest share of nonpracticing entities.[36] I broke down the nationalities of the litigants at each court and found strong concentrations of patent holders from certain countries at the courts (cf. Table 2.12 in the Appendix). For instance, Japanese patent holders file their action almost

---

[36]Nonpracticing entities are primarily active in the field of electrical engineering technologies.

exclusively in Düsseldorf, while the majority of Italian patent holders go to Mannheim.

Patents subject to an infringement proceeding in Düsseldorf appear to be the most international with the highest share of *EP* bundle patents and PCT filings. While I found that patents in Mannheim are cited relatively more frequently in the first five years, I did not observe any significant difference in backward citations to patents. Citations to nonpatent literature are more prevalent for patents in Mannheim. Equally, the number of claims, assigned IPC subclasses, and equivalents is highest for patents in Mannheim. The average length of examination is longest for patents in Düsseldorf, even though an acceleration of the examination was requested in almost 20% of cases. In contrast, patent age shows little variance across the three courts. Examining the invalidation history of the patents, I found that patents in Mannheim have faced the least prior oppositions but by far the most prior revocation proceedings. Still, these validity challenges ended mostly in withdrawals, so an equally low share of patents can be considered solidified at each court.

The distribution of technology areas is quite remarkable. The regional court in Mannheim predominantly hears disputes on patents in the field of electrical engineering, while Düsseldorf hears the most cases on chemistry patents. I plotted the litigants with German residency on a country map to visualize the spatial distribution by court and technology main area (cf. Figures 2.9 and 2.10 in the Appendix). Although the concentration of patent holders and alleged infringers in west and southwest Germany mirrors the location of important industrial regions, neither the existence nor the proximity of industrial clusters can adequately explain the number of cases and the disproportionate representation of technology areas at the three regional courts. For instance, the Mannheim regional court sees very few cases on chemistry patents, even though the surrounding *Rhein-Neckar-Gebiet* is host to a large cluster of chemical firms.

2.6   Empirical Model and Results

2.6.1   Empirical Model

Modeling court selection

The primary goal of the econometric model of court selection is to recover estimates of the plaintiffs' preferences. To do so, I draw on an additive random-utility model. For the plaintiff of dispute i and chosen court j among J courts with $J = \{DU, MA, MU\}$, the utility U_{ij} is the sum of deterministic component V_{ij} and the unobserved random term ϵ_{ij}. I observe the outcome $y_i = j$ if $U_{ij} > U_{ik} \quad \forall \quad k \neq j$. Accordingly, I specify the alternative-specific conditional logit model with the error terms as *iid* extreme value distributed:

$$U_{ij} = \beta X'_{ij} + \gamma_j Z'_i + \epsilon_{ij} \tag{2.6.1}$$

so that

$$p_{ij} = \Pr(y_i = j) = \frac{\exp\left(\beta X'_{ij} + \gamma_j Z'_i\right)}{\sum_{k=1}^{n} \exp\left(\beta X'_{ij} + \gamma_j Z'_i\right)} \quad \text{with} \quad j = DU,\ MA,\ MU,$$

where the vector X'_{ij} represents alternative-specific regressors and vector Z'_i covers case-specific regressors. The alternative-specific regressors X'_{ij} include the opportunity costs OPC_{ij}, and the distance to court of both the plaintiff, TCP_{ij}, and the defendant, TCD_{ij}. The case-specific regressors Z'_i include year, technology and litigant (size and residence) controls as well as dummies indicating a multijurisdictional dispute and whether the plaintiff had any prior experience with the court.

The plaintiff of dispute i goes to court y_i that corresponds to the highest value function, i.e.,

$$y_i = \arg \max_{j \in \{DU,\ MA,\ MU\}} U_{ij}.$$

Using cross-sectional data with each observation representing a dispute, the main empirical specification is as follows:

$$U_{ij} = \beta_1 OPC_{ij} + \beta_2 TCP_{ij} + \beta_3 TCD_{ij} + \gamma_{1j} PRIOR_i + \gamma_{2j} STAKE_i +$$
$$+ \gamma_{3j} TOP_i + \gamma_{4j} MULTI_i + \gamma_{5j} TECH_i + \gamma_{6j} LIT_i + \gamma_{7j} YEAR_i + \epsilon_{ij}. \tag{2.6.2}$$

Following from the theoretical model, I anticipate that $\beta_1 < 0$ and $\beta_2 < 0$; i.e., an increase in the plaintiff's opportunity and transaction costs at one court increases the probability of selecting an alternative court. If I find that $\gamma_{2j} \neq \gamma_{2k}\ \exists\ k \neq j$, I can infer that plaintiffs assume court-specific probabilities to win their case. For instance, if $\gamma_{2j} < \gamma_{2k}$, the plaintiffs perceive court j as less patent-friendly than court k. Thus, the higher the value at stake (*STAKE*), which is the sum of the litigation value and all legal costs, the more likely a plaintiff is to select the alternative court k, where the expected liability of infringement is higher.

I estimate the model on a sample of all infringement proceedings based on patents in force and without a prior preliminary injunction. I also regress court selection on several subsamples defined by the plaintiffs' size and product market proximity to show robustness. Since one of the regressors, the opportunity costs OPC_{ij}, is a function of an endogenous variable, the predicted length of proceeding, I correct the standard errors through bootstrapping.[37]

Predicting the ex ante expected length of proceeding

Prior to estimating the plaintiff's court selection, I first need to estimate his opportunity costs due to delay in judgment. I construct the alternative-specific measure for opportunity costs, OPC_{ij}, in line with the theoretical model in Section 2.2.1. For the sake of simplicity, I consider opportunity costs a direct function of the litigation value L (cf. Section 2.8.4 in the Appendix for an elaborate derivation). The opportunity costs are operationalized as

$$OPC_{ij} = \alpha_i \frac{L_i\left((1-\delta_i)^{l_{ij}}-1\right)}{(1-\delta_i)^T-1}. \tag{2.6.3}$$

Unfortunately, the ex ante expected length of proceeding l_{ij}, which determines the amount of opportunity costs, remains unobserved in the data for multiple reasons. First, I only observe the actual length of proceeding at the selected court, not at alternative courts. Second, most proceedings end prematurely in settlement. Third, the actual length of the proceeding is chamber-specific. This length diverges from the court-specific ex ante expected length if the court has two chambers and the respective judges differ, e.g., in terms of expertise.[38] For all these reasons, I am left with an unclear picture of the time the plaintiff initially expects a particular court to take until judgment.

I make use of the fact that the German patent infringement main proceeding is characterized by a very structured process that offers *de facto* only two discrete events as cause for considerable delay (cf. Section 2.3.2). I construct the ex ante expected length of proceeding until judgment at each court j for each case i with the following equation:

$$\widehat{l}_{ij} = \bar{l}_j^{ord} + \Pr_{ij}(\widehat{\text{stay}=1}|\text{inv. proc.}=1) \cdot l^{stay} + \Pr_{ij}(\widehat{\text{expert}}=1) \cdot \bar{l}^{expert} - \\ - \Pr_{ij}(\widehat{\text{stay}=1}|\text{inv. proc.}=1) \cdot \Pr_{ij}(\widehat{\text{expert}}=1) \cdot \min\left(l^{stay}, \bar{l}^{expert}\right). \tag{2.6.4}$$

This equation makes the following simplifying assumptions: the expected length of any ordinary proceeding is the median length of all proceedings with judgment at a specific court in a given year (year indices omitted). Further, the probabilities of the two delaying events, namely stay and expert opinion, are predicted based on the results of parametric models estimating the probabilities of either event (cf. Table 2.6 for an overview). Here again, the dependent

---

[37]I drew 133 random samples with replacement equal in size to the original sample.
[38]As noted earlier, the plaintiff cannot foresee which chamber his case will be assigned to.

variables, i.e., the delaying events, are incompletely observed, because many proceedings end in settlement prior to the main oral hearing, which usually reveals how the court decides on a stay and the need for an expert opinion.

The most straightforward way would be to estimate the factors determining the length of proceedings directly; however, this approach reduces the estimation sample to proceedings with judgment.[39] By estimating the probability that delaying events will occur, I can extend the sample to all proceedings that did not end in a settlement prior to the main oral hearing.[40] This approach increases the sample size considerably while minimizing selection problems.

Besides the probabilities of the two delaying events, I also need to define how long a judgment will be deferred if the court decides to stay the proceeding (l^{stay}) or request an expert opinion ($\bar{l}^{expert}$). The general term l^{stay} can be either dispute-specific, l_{ij}^{stay}, if an invalidity proceeding is already pending at the time of the plaintiff's court selection, or a general estimation, $\bar{l}_j^{stay}$, if no invalidity action is pending at the time of the plaintiff's court selection. The length of the stay $\bar{l}_j^{stay}$ is determined exogenously as the difference between the median lengths of all first instance revocation/opposition proceedings filed in that year plus the observed time difference (median time difference) between the filing of the invalidity proceeding and the filing of the infringement proceeding Δ_i ($\bar{\Delta}$) plus the court-specific expected length of the ordinary proceeding $\bar{l}_j^{ord}$.[41] I fix the delay caused by an expert opinion, $\bar{l}^{expert}$, at twenty-four months across all three courts, which is in line with estimates by Kühnen (2012) and interviewed practitioners.

I estimate the probability for each event independently. Certain characteristics associated with the patent in dispute (PAT_i), the litigants (LIT_i), the proceeding ($PROC_i$), and the judges (JUD_j), comprise the regressors in the estimations. Table 2.6 gives an overview of which characteristics are used in which model to regress the respective delaying event.

I predict the probability of the request for an expert opinion with the following probit model:

$$\Pr(\text{expert} = 1) = \Phi(\beta_1^{expert} PAT^{expert} + \beta_2^{expert} JUD^{expert} +$$
$$+ \beta_3^{expert} PROC^{expert}). \tag{2.6.5}$$

---

[39]Settlements may occur at any time during the proceeding; for instance, I also observe settlements months after the main oral hearing if a stay or an expert opinion occurred (cf. Figure 2.2).

[40]As the data were collected directly from the court case files with transcripts of the main hearing and written court orders, I have data on the court's decision to stay or to call in an expert even if the proceeding ended in a settlement shortly after.

[41]For a year-by-year overview of $\bar{l}_j^{stay}$, see Table 2.13 in the Appendix.

Table 2.6: Estimation models for delaying events in infringement proceeding

Model	Expert opinion (E) Probit	Stay of proceeding (S) Probit	Probit with sample selection[a]	
Dependent variable				
Discrete event	Expert opinion	Stay	Stay	Invalidity proceeding
Independent vectors/variables				
Patent characteristics (PAT_i)				
Patent intricacy	✓			
Patent quality		✓	✓	✓
Patent value				✓
Patent technology	✓	✓	✓	✓
Litigant characteristics (LIT_i)				
Size of litigants				✓
Residence of litigants				✓
Legal representation				✓
Judicial characteristics (JUD_i)				
Judicial expertise	✓			
Judge-/court-specific idiosyncrasy	✓	✓	✓	✓
Procedural characteristics $(PROC_i)$				
Litigation value	✓	✓	✓	✓
Multinational litigation	✓	✓	✓	✓
Timing of inv. proc.		✓	✓	
Decision on noninfringement		✓	✓	
Caused delay[b]	$\bar{l}^{expert}$	$\bar{l}_t^{inv} + \Delta_i - \bar{l}_{jt}^{ord} = l_{ijt}^{stay}$	$\bar{l}_t^{inv} + \bar{\Delta} - \bar{l}_{jt}^{ord} = \bar{l}_{jt}^{stay}$	

[a] For cases with no invalidity proceeding pending at time of court selection.
[b] Ex ante likelihood of stay of proceeding predicted on basis of average delay of invalidity proceeding $\bar{\Delta}$ and average rate of decision on noninfringement (conditional on no settlement). Values used for the median lengths $\bar{l}$ can be found in Table 2.13 in the Appendix.

As the stay of a proceeding is conditional on a parallel invalidity proceeding, I need to distinguish between two cases: whether an invalidity proceeding is already pending at the time of the plaintiff's court selection or not. In the latter case the probability of a stay is predicted by a probit model with sample selection (Van de Ven and Van Praag, 1981). I specify the selection equation as

$$\Pr(\text{inv. proc.} = 1) = \Phi(\beta_1^{inv} PAT^{inv} + \beta_2^{inv} LIT^{inv} +$$
$$+ \beta_3^{inv} JUD^{inv} + \beta_4^{inv} PROC^{inv}), \qquad (2.6.6)$$

and the resultant binary outcome equation as

$$\Pr(\text{stay} = 1 | \text{inv. proc.} = 1) = \Phi(\beta_1^{stay} PAT^{stay} + \beta_2^{stay} JUD^{stay} +$$
$$+ \beta_3^{stay} PROC^{stay}). \qquad (2.6.7)$$

I draw on characteristics of the patent and the litigants as exclusion restrictions. That is, I include the corresponding variables in the selection equation, but exclude them from the binary outcome equation. The value of the patent as well as the size and residence of the litigants should play a role in the likelihood of the filing of an invalidity proceeding, but should not have any effect on the decision of the court to actually grant a stay. As the event of an accelerated examination request serves as a sole patent value indicator (cf. Table 2.2), I particularly focus on this variable as an exclusion restriction.

When regressing the decision to request an expert opinion, I can reject equal propensities among the courts (cf. Table 2.7). With the Düsseldorf regional court as baseline, I observe a positive effect of the remaining two courts on the likelihood of an expert opinion. Patent intricacy, captured by the count of nonpatent literature backward citations and IPC subclasses, shows a significant positive effect as well. The constructed measures of the judges' expertise appear significant. I turn to judge controls in the fourth specification (E4). These controls and their technology interaction terms are partly significant (cf. Table 2.14 in the Appendix). In particular, electrical engineering patents appear less likely to be subject to expert opinions at the Mannheim regional court, which hears a disproportional share of cases in this technology area. I can conclude that the regional courts in Mannheim and Munich are in general more likely to request expert opinions than the Düsseldorf regional court, with Munich even more likely to do so than Mannheim, while the magnitude differs at a technology level.

Table 2.8 shows the results of regressing the decision to grant a stay of the infringement proceeding conditional on a parallel invalidity proceeding. The assumption that the value of the patent has no effect on the judge's tendency to grant a stay is supported by the nonsignificant effect of litigation value on the grant of a stay. This supports my decision to use sole patent value indicators as an exclusion restriction. The results of the selection equation show that the identity of the judge hearing the infringement proceeding has little effect on the likelihood of a validity challenge. In contrast, the results of the outcome equation show highly significant judge effects. I find that the Mannheim and Munich regional courts are more likely to stay a proceeding compared to Düsseldorf. This result may be due to the differing ordinary lengths of proceeding; i.e., the regional court in Düsseldorf stays the proceeding less often because a

Table 2.7: Probit model results: incidence of expert opinion

	(E1) Expert opinion	(E2) Expert opinion	(E3) Expert opinion	(E4) Expert opinion
Court effects				
LG Mannheim (d)	0.136***	0.113***	0.342***	
	(0.031)	(0.026)	(0.094)	
LG Munich (d)	0.243***	0.238***	0.652***	
	(0.050)	(0.049)	(0.117)	
Patent intricacy				
Nonpatent literature ratio		0.073	0.074	0.082*
		(0.040)	(0.042)	(0.041)
No. of claims		0.001	0.001	0.001
		(0.001)	(0.001)	(0.001)
IPC subclasses count		0.014**	0.015**	0.015**
		(0.004)	(0.005)	(0.005)
Age of patent		−0.003	−0.002	−0.003
		(0.002)	(0.002)	(0.002)
Grant lag (difference from mean in days)		0.000	−0.000	−0.000
		(0.000)	(0.000)	(0.000)
Prior infringement decision on patent (d)			−0.038	−0.033
			(0.027)	(0.027)
Procedural characteristics				
Litigation value (in thousand €, log)	0.004	0.007	0.008	0.005
	(0.010)	(0.009)	(0.008)	(0.008)
Multijurisdictional litigation (d)			0.054	0.041
			(0.061)	(0.059)
Judicial expertise				
Tenure as judge (in years)			−0.008***	0.012
			(0.002)	(0.007)
Prior exposure to technology area			0.068***	0.037
			(0.019)	(0.048)
Judge effects	No	No	No	Yes***
Technology effects	No	No	Yes	Yes
Court x Technology effects	No	No	Yes	No
Judge x Technology effects	No	No	No	Yes
Year effects	Yes	Yes	Yes*	Yes
Pseudo R^2	0.047	0.062	0.080	0.103
Observations	1,690	1,690	1,690	1,645

Marginal effects; Standard errors in parentheses
(d) for discrete change of dummy variable from 0 to 1
* $p < 0.05$, ** $p < 0.01$, *** $p < 0.001$

Notes: The sample consists of all patent infringement proceedings based on patents in force that did not end in settlement prior to the call for an expert opinion ($\bar{l}_{jt}^{ord} - 50$ days). Proceedings with judgment by consent/default decree excluded. The unit of observation is at the case level. 45 observations in regression (E4) dropped due to perfectly predicted failure. Base line regional court: LG Düsseldorf (DU). Judge effects, technology effects, and judge x technology interaction effects displayed in Table 2.14 in the Appendix. Standard errors clustered by patent.

decision on the invalidity proceeding is more likely to be available in time. I therefore control for the lag of the invalidity proceeding relative to the court-specific length of an ordinary proceeding and find the judge effects remain robust.

Having identified the determinants of the likelihood of delaying events, I am now able to predict the expected length of proceeding for each case at each court. I adjust judge-specific variables to court-specific variables where necessary, and impute ex ante unknown variables (lag of invalidity proceeding and the decision on noninfringement) with average values. I then perform out-of-sample predictions of the likelihood for delaying events for all cases at each court (cf. Table 2.9). With these values I calculate the ex ante expected lengths of proceeding according to equation (2.6.4). The densities of the court-specific lengths can be found in Figure 2.3 for all proceedings and in Figure 2.11 in the Appendix broken down by technology area.[42]

The densities differ among the regional courts; however, in contrast to Figure 2.2, the Düsseldorf regional court now appears to be faster than the Mannheim and Munich regional courts. Although the Mannheim regional court may still be faster than the regional court Düsseldorf, the majority of cases seem to take considerably longer. This is because the probability of delaying events weighs heavier on the ex ante expected length of proceedings in Mannheim and Munich than in Düsseldorf for most cases within the overall population.

---

[42]I also compare the densities of both conditional and unconditional samples in Figure 2.12 in the Appendix.

Table 2.8: Probit model with sample selection results: incidence of stay of proceeding

	(S1)		(S2)		(S3)		(S4)		(S5)	
	INF stayed	INV filed	INF stayed	INV filed	INF stayed	INV filed	INF stayed	INV filed	INF stayed	INV filed
Procedural characteristics										
Decision on noninfringement (d)	-0.956***		-0.914***		-0.984***		-0.984***		-0.984***	
	(0.160)		(0.174)		(0.154)		(0.153)		(0.153)	
Lag of invalidity proc. (in months)							0.003		0.003	
							(0.003)		(0.003)	
Litigation value (in thousand €, log)	0.115	0.252***	0.125	0.298***	0.057	0.201***	0.056	0.201***	0.056	0.201***
	(0.063)	(0.039)	(0.065)	(0.040)	(0.062)	(0.042)	(0.062)	(0.042)	(0.062)	(0.042)
Multijurisdictional litigation (d)	0.742***	1.107***	0.780***	1.054**	0.668**	0.933***	0.674**	0.934**	0.674**	0.934**
	(0.213)	(0.284)	(0.218)	(0.327)	(0.231)	(0.293)	(0.232)	(0.293)	(0.232)	(0.293)
Judge effects										
Judge 1 (LG DU) (d)	0.626*	-0.027	0.604*	-0.093	0.687*	0.038	0.655*	0.038	0.655*	0.038
	(0.276)	(0.240)	(0.276)	(0.214)	(0.319)	(0.224)	(0.320)	(0.224)	(0.320)	(0.224)
Judge 2 (LG DU) (d)	0.742**	0.059	0.727**	-0.033	0.846*	0.067	0.809*	0.067	0.809*	0.067
	(0.279)	(0.241)	(0.282)	(0.214)	(0.341)	(0.236)	(0.340)	(0.236)	(0.340)	(0.236)
Judge 1 (LG MA) (d)	1.265**	-0.064	1.241**	-0.106	1.337**	0.137	1.308**	0.137	1.308**	0.137
	(0.393)	(0.304)	(0.391)	(0.287)	(0.416)	(0.289)	(0.416)	(0.289)	(0.416)	(0.289)
Judge 2 (LG MA) (d)	1.385***	0.290	1.356***	0.117	1.485***	0.305	1.447***	0.306	1.447***	0.306
	(0.301)	(0.253)	(0.317)	(0.236)	(0.372)	(0.262)	(0.373)	(0.262)	(0.373)	(0.262)
Judge 3 (LG MA) (d)	1.154***	0.328	1.184***	0.233	1.245***	0.374	1.208***	0.374	1.208***	0.374
	(0.317)	(0.272)	(0.322)	(0.252)	(0.351)	(0.264)	(0.351)	(0.264)	(0.351)	(0.264)
Judge 1 (LG MU) (d)	1.099**	0.045	1.049**	0.042	1.161**	0.216	1.126**	0.216	1.126**	0.216
	(0.362)	(0.293)	(0.365)	(0.273)	(0.418)	(0.290)	(0.417)	(0.290)	(0.417)	(0.290)
Judge 2 (LG MU) (d)	1.479***	-0.258	1.371**	-0.372	1.631***	-0.074	1.620***	-0.074	1.620***	-0.074
	(0.424)	(0.305)	(0.434)	(0.289)	(0.446)	(0.301)	(0.445)	(0.301)	(0.445)	(0.301)
Exclusion restriction										
Accelerated examination requested (d)		0.547***		0.333**		0.284*		0.284*		0.284*
		(0.103)		(0.108)		(0.114)		(0.114)		(0.114)
Patent characteristics	No	No	No	No	Yes***	Yes***	Yes***	Yes**	Yes***	Yes***
Invalidation history	No	No	No	No	Yes**	Yes**	Yes**	Yes**	Yes**	Yes**
Litigant characteristics	No	No	No	No	No	No	No	Yes***	No	Yes***
Technology effects	No	No	No	No	Yes	Yes	Yes	Yes	Yes	Yes
Year effects	No	No	No	No	Yes	No	Yes	Yes	Yes*	Yes*
Observations	1617		1617		1617		1617		1617	

Marginal effects; Standard errors in parentheses; (d) for discrete change of dummy variable from 0 to 1; * $p < 0.05$, ** $p < 0.01$, *** $p < 0.001$

Notes: INF: infringement proceeding. INV: invalidity proceeding. The sample consists of all patent infringement proceedings based on patents in force that did not end in settlement prior to the decision on a stay of proceeding ($J_{jt}^{inf} - 50$ days). Proceedings with judgment by consent/default decree excluded. Proceedings based on patent *EP0402973* ($N > 50$) excluded. The unit of observation is at the case level. Base line regional court: LG Düsseldorf (DU). Base line judge: Judge 3 (LG DU). Patent characteristics: Forward citations in first 5 years, backward citations, nonpatent backward citations (ratio), no. of claims, age of patent, age of patent (squared), IPC count, *EP* (d), *PCT* (d), family size, grant lag (difference from mean in days). Invalidation history: Patent solidified through opp. proc. (d), patent challenged through rev. proc. (d), patent solidified through rev. proc. (d). Litigant characteristics: Size categories (d), background categories (d), top legal representative (d). Standard errors clustered by patent.

Table 2.9: Predicted likelihoods of delaying events by court and technology main area

	Expert opinion		Validity challenge		Stay	
	Mean	Std. dev.	Mean	Std. dev.	Mean	Std. dev.
Electrical engineering						
LG DU	0.19	0.12	0.50	0.33	0.33	0.16
LG MA	0.25	0.12	0.56	0.32	0.57	0.18
LG MU	0.42	0.10	0.51	0.33	0.59	0.18
Instruments						
LG DU	0.07	0.05	0.65	0.31	0.22	0.12
LG MA	0.30	0.13	0.70	0.29	0.43	0.17
LG MU	0.33	0.09	0.65	0.31	0.45	0.16
Chemistry						
LG DU	0.09	0.04	0.69	0.32	0.25	0.13
LG MA	0.29	0.12	0.74	0.29	0.47	0.16
LG MU	0.27	0.10	0.69	0.32	0.49	0.16
Mechanical engineering						
LG DU	0.11	0.05	0.57	0.30	0.20	0.10
LG MA	0.19	0.05	0.64	0.28	0.41	0.15
LG MU	0.33	0.07	0.58	0.30	0.42	0.15
Other						
LG DU	0.08	0.04	0.52	0.30	0.19	0.09
LG MA	0.23	0.06	0.59	0.28	0.40	0.14
LG MU	0.27	0.07	0.53	0.30	0.41	0.14
All	0.23	0.08	0.59	0.31	0.40	0.16

Notes: The sample consists of all patent infringement proceedings based on patents in force. The unit of observation is at the case level. The results reflect out-of-sample predictions; that is, predictions are on all proceedings for each court. Predicted likelihoods of suspension conditional on a parallel validity challenge.

2.6.2 Results

Table 2.10 shows the results of the alternative-specific conditional logit model on court selection. In regressions (C1) to (C3) I use alternative measures for opportunity cost due to delay in judgment with increasing similarity to the preferred operationalization presented in regression (C4). I assume the longer the time until judgment at one court, the more likely an alternative court will be chosen. Interacting expected length with litigation value leads to a low significant effect (cf. regression (C1)). Taking the maximum length of patent protection into account makes regression (C2) significant again. Regression (C3) introduces technology-specific discount factors but still ignores the product market proximity of the litigants. The effect remains negative while again losing some significance. In regression (C4) I finally present the results of the preferred specification with opportunity costs weighted by product market proximity. Here

Figure 2.3: Ex ante predicted lengths of infringement proceedings until judgment by court (densities)

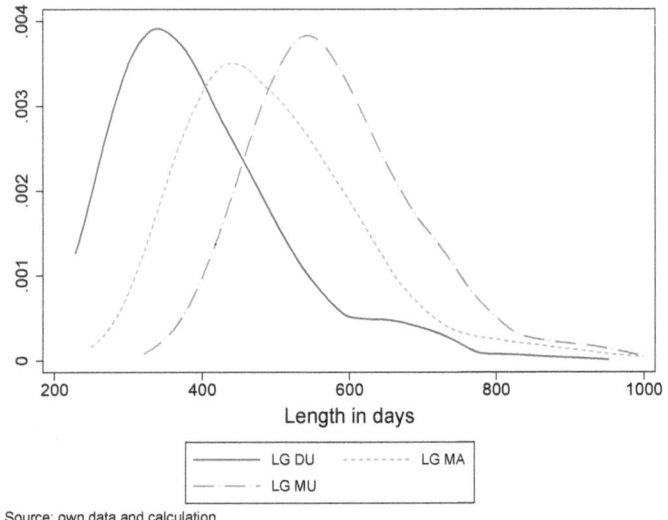

Source: own data and calculation

Notes: The unit of observation is the infringement proceeding. Truncated at 1,000 days.

again, the effect of opportunity costs is significant and negative. Knowing that the approximation of product market proximity is rather crude, I use the former measure from (C3) on a sample of litigants with low product market proximity ($\alpha \leq 0.5$) in regression (C5) and on a sample of litigants with high product market proximity ($\alpha > 0.5$) in regression (C6). Since the effect of opportunity costs is insignificant in the former regression (C5), but highly significant in the latter regression (C6), I consider the theoretical predictions confirmed: delay in judgment matters most for plaintiffs that operate in the same product market as the defendant.

For all regressions I observe a robust highly significant negative effect of the plaintiff's transaction costs on court selection. The samples used in regressions (C7) and (C8) differ in size of the plaintiffs. I find that the effect of transaction costs is almost twice as large for small plaintiffs in regression (C7) as for large plaintiffs in regression (C8). These results are in line with the theoretical predictions that costs weigh heavier on small, financially constrained, litigants and therefore play a larger role in their court choice.

I also observe a negative effect of the defendant's transaction costs. However, I refrain from interpreting this result as evidence for their role in the plaintiff's consideration for court choice. This effect could be a corollary of the fact that the defendant's residence fulfills the most basic requirements for territorial jurisdiction (cf. Section 2.3.1).[43] Since I find only negligible correlation between the plaintiff's and the defendant's transaction costs ($r = 0.029$), I rule out an interrelationship between the two effects.

Looking at the choice-specific variables, I find that representation by a top law firm has a negative effect on the selection of the Mannheim and Munich regional courts.[44] Multijurisdictional disputes between the litigants have no significant effect on court selection. I find no significant effect of the value at stake on court selection at the Mannheim regional court, but a somewhat significant negative effect in regression (C4) at the Munich regional court. Due to the lack of robustness across the regressions, I am reluctant to confirm a perceived anti-patentee bias associated with the regional court Munich. Further, the effect may also originate from interactions with other court-specific factors at the regional court in Munich which I am not able to observe. These may include a higher likelihood of appeal or a more austere calculation of damages. In a similar vein, prior experience with the Mannheim regional court shows no substantial effect on selecting the court again. In contrast, prior experience with the Munich regional court appears to have a strong negative effect on the decision to revisit the court.[45]

---

[43]Apparently, the requirements for territorial jurisdiction at all courts are not met in all cases within the sample, even though I rigorously excluded proceedings with a prior preliminary injunction or trade fair context.

[44]This effect is robust for an alternative classification of legal representatives (not shown).

[45]In further specifications (not shown), I additionally control for the outcome of the prior proceedings, but do not find any effect.

Table 2.10: Alternative-specific conditional logit model results: court selection

	(C1) Court	(C2) Court	(C3) Court	(C4) Court	(C5) Court	(C6) Court	(C7) Court	(C8) Court
Alternative-specific variables								
Expected length x litigation value	-0.002* (0.001)							
Opportunity costs ($\alpha = 1$, $\delta = 0$)		-0.015** (0.005)						
Opportunity costs ($\alpha = 1$)			-0.015* (0.007)					
Opportunity costs				-0.012* (0.005)	-0.013 (0.010)	-0.023** (0.008)	-0.026 (0.012)	-0.010* (0.004)
Case-specific variables: LG Mannheim (MA)								
Plaintiff distance to court (in thousand km, log)	-0.482*** (0.058)	-0.496*** (0.059)	-0.489*** (0.058)	-0.490*** (0.057)	-0.571*** (0.101)	-0.443*** (0.076)	-0.713*** (0.122)	-0.429*** (0.071)
Defendant distance to court (in thousand km, log)	-0.297*** (0.048)	-0.306*** (0.047)	-0.308*** (0.047)	-0.303*** (0.046)	-0.313*** (0.076)	-0.286*** (0.063)	-0.326*** (0.094)	-0.295*** (0.062)
Prior case at court (in last 3 years) (d)	0.043 (0.179)	0.075 (0.181)	0.100 (0.182)	0.021 (0.171)	-0.089 (0.296)	0.179 (0.209)	-0.561 (0.508)	0.160 (0.184)
Value at stake (in million €)	0.017 (0.164)	0.080 (0.158)	0.099 (0.176)	-0.153 (0.087)	-0.006 (0.248)	0.792*** (0.273)	-0.665* (0.298)	-0.048 (0.091)
Multijurisdictional litigation (d)	0.461 (0.461)	0.525 (0.483)	0.400 (0.465)	0.478 (0.487)	0.343 (2.509)	0.246 (0.165)	3.741* (1.827)	-0.565 (3.030)
Top legal representative (JUVE) (d)	-2.150*** (0.161)	-2.019*** (0.152)	-2.063*** (0.158)	-2.181*** (0.142)	-2.255*** (0.345)	-1.844*** (0.197)	-1.708*** (0.298)	-2.304*** (0.178)
Case-specific variables: LG Munich (MU)								
Prior case at court (in last 3 years) (d)	-1.122* (0.448)	-1.196** (0.448)	-1.136* (0.449)	-1.176*** (0.450)	-0.697 (2.056)	-1.467*** (2.710)	-0.690 (7.276)	-1.263 (4.823)
Value at stake (in million €)	-0.282 (0.409)	-0.459 (0.420)	-0.444* (0.476)	-0.679* (0.594)	0.031 (1.414)	-0.093 (0.812)	-0.902 (0.812)	-0.521 (2.530)
Multijurisdictional litigation (d)	0.399 (2.711)	0.569 (3.039)	0.379 (2.674)	0.420 (2.696)	-14.512*** (12.402)	1.415* (0.703)	2.573** (0.882)	0.644 (1.580)
Top legal representative (JUVE) (d)	-1.660*** (0.240)	-1.562*** (0.243)	-1.594*** (0.243)	-1.669*** (0.244)	-1.440 (1.504)	-1.478** (0.551)	-1.377 (4.091)	-1.611** (0.416)
Plaintiff characteristics	Yes	Yes	Yes	Yes	Yes	Yes	Yes	Yes
Technology effects	Yes**	Yes**	Yes**	Yes**	Yes**	Yes	Yes*	Yes**
Year effects	Yes**	Yes***	Yes***	Yes***	Yes**	Yes**	Yes*	Yes***
Pseudo R^2	0.325	0.353	0.343	0.330	0.427	0.279	0.303	0.361
Observations	6,333	6,333	6,333	6,333	2,895	3,438	1,308	5,025

Marginal effects; Bootstrapped standard errors in parentheses; (d) for discrete change of dummy variable from 0 to 1; * $p < 0.05$, ** $p < 0.01$, *** $p < 0.001$

Notes: The sample consists of all patent infringement proceedings based on patents in force without a (prior) request for a preliminary injunction. The unit of observation is at the case level. Base line technology area: Other. Base line regional court: LG Düsseldorf (DU). Plaintiff characteristics: Size categories (d), residence categories (by country) (d). Bootstrapped standard errors reported. C5: baseline sample with $\alpha \leq 0.5$. C6: baseline sample with $\alpha > 0.5$. C7: baseline sample with micro and small plaintiffs. C8: baseline sample with medium and large plaintiffs.

2.7 Conclusion

This study analyzed the determinants of court selection in patent litigation using data from German regional courts. Several results from this study may contribute to both theoretical and practical discussions on the optimal designs of patent litigation systems – most notably the ongoing debate regarding the rules and structure of the Unified Patent Court (*UPC*).

I found that the regional courts in Düsseldorf, Mannheim and Munich differ in their ordinary length of proceeding and likelihood of delaying events. The primary results reveal that plaintiffs consider their economic loss from delayed judgment when choosing their court. Speedy enforcement is highly valued, especially if the litigants operate in the same product market. The self-allocation of patent disputes among multiple courts leads to efficiency gains, because courts have an incentive to invest in specialization, accrue experience and induce procedural innovations to attract suitable patent disputes. The decision to divide the responsibilities of the UPC's regional divisions in Paris, London, and Munich by technology field may facilitate the accumulation of relevant technical expertise at the respective courts to the benefit of faster and more coherent judgments.

The distance to a particular court has a negative effect on the plaintiff's court selection. Here, the magnitude of the effect is considerably larger for small plaintiffs. Since small plaintiffs highly value local access to court, they may be reluctant to file their action in the first place due to overwhelming transaction costs in judicial systems that force them to seek judicial relief at a distant court. The Unified Patent Court will consist of several spatially dispersed local and regional divisions all over Europe. While this ensures that plaintiffs gain local access to legal patent enforcement, the proceeding can be transferred to the central division (in Paris), for instance, in the case of a validity challenge. This again may increase the plaintiff's transaction costs.

I further find evidence that courts do not show perfect uniformity in their decision making. The Düsseldorf regional court tends to grant a stay of proceeding less often than the other two courts and plaintiffs perceive the Munich regional court as having a stronger anti-patentee bias. Prior research has shown that lack of uniformity increases legal uncertainty and impedes settlements (cf. Galasso and Schankerman, 2010). While a centralized appeals court at the Unified Patent Court may promote uniformity in judicial decision making, judges maintain considerable discretion in case management and procedural options prior to a judgment on the merits.

2.8   Appendix to Chapter 2

2.8.1   Figures

Figure 2.4: Court structure in Germany's patent system (Cremers *et al.*, 2013, amended)

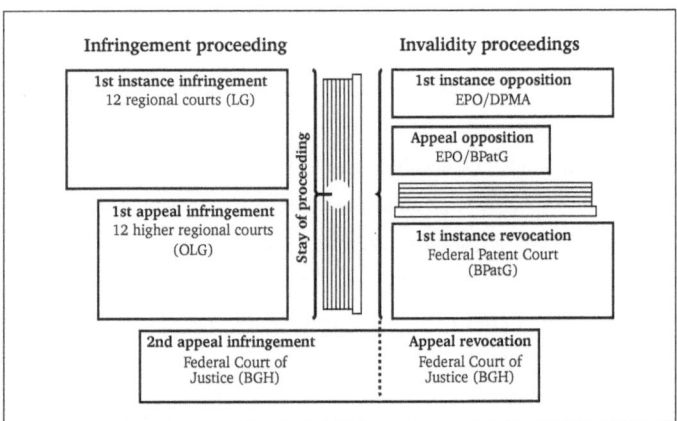

Figure 2.5: Proceedings linked to patent infringement disputes (own illustration)

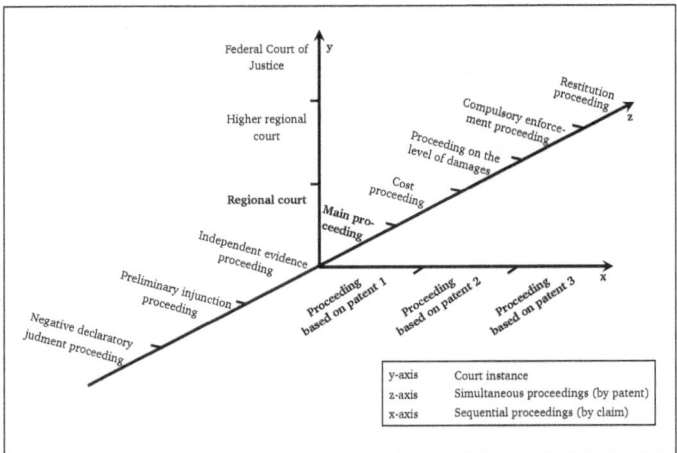

Figure 2.6: Structure of the infringement main proceeding (own illustration)

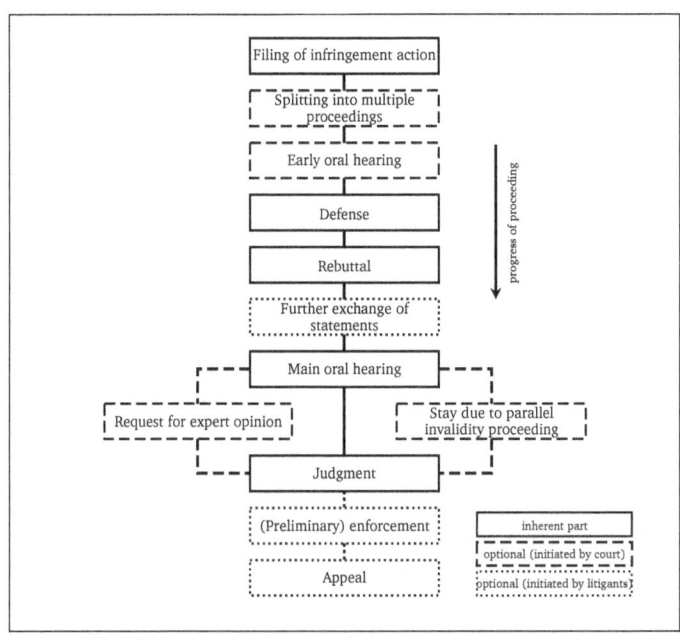

Figure 2.7: Length of main infringement proceedings with judgment by court and year

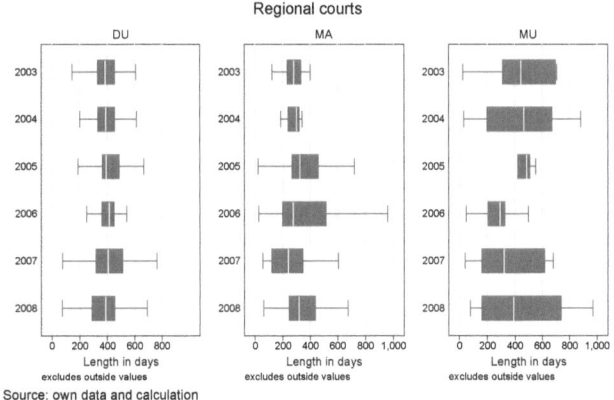

Notes: The sample consists of all patent infringement proceedings based on patents in force without a (prior) request for a preliminary injunction. The unit of observation is at the case level. Truncated at 1,000 days.

Figure 2.8: Length of main infringement proceedings with settlement by court and year

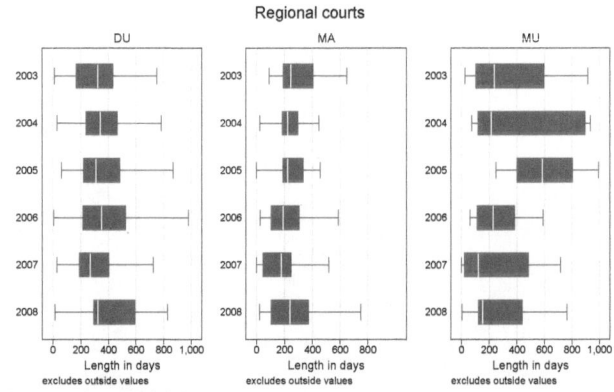

Source: own data and calculation

Notes: The sample consists of all patent infringement proceedings based on patents in force without a (prior) request for a preliminary injunction. The unit of observation is at the case level. Truncated at 1,000 days.

Figure 2.9: Spatial distribution of patent holders by court and technology main area

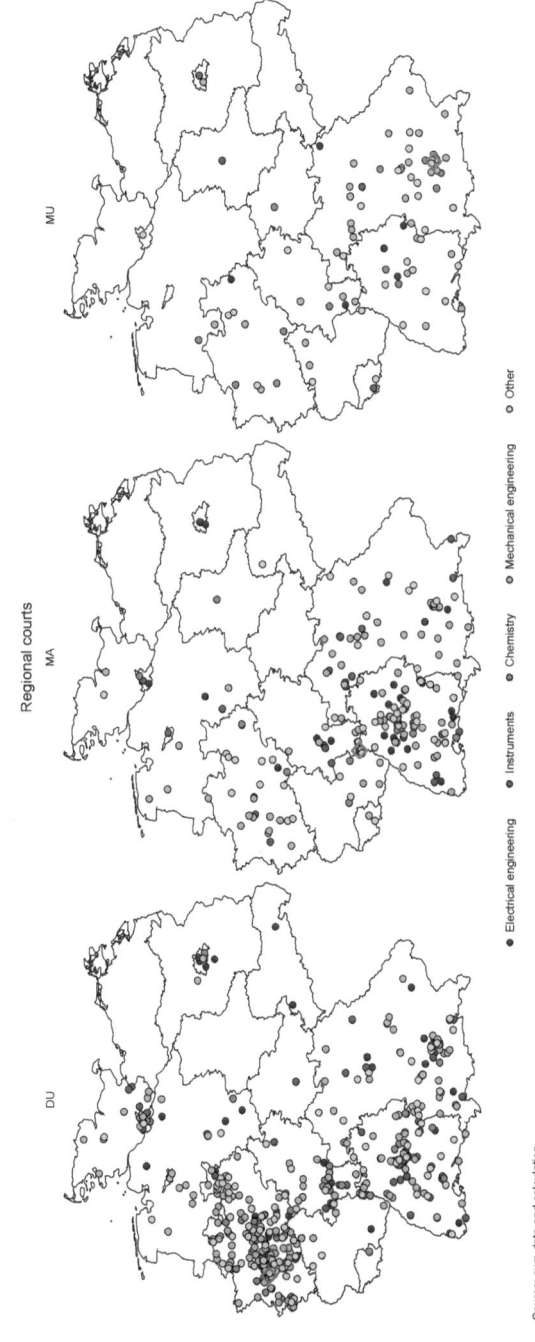

Regional courts

DU   MA   MIU

● Electrical engineering   ● Instruments   ● Chemistry   ● Mechanical engineering   ○ Other

Source: own data and calculation

Notes: The sample consists of all patent infringement proceedings based on patents in force. The unit of observation is at the case level. Only patent holders with residence in Germany represented. In case of multiple patent holders, the one closest to court is chosen. Diamond symbolizes place of regional court.

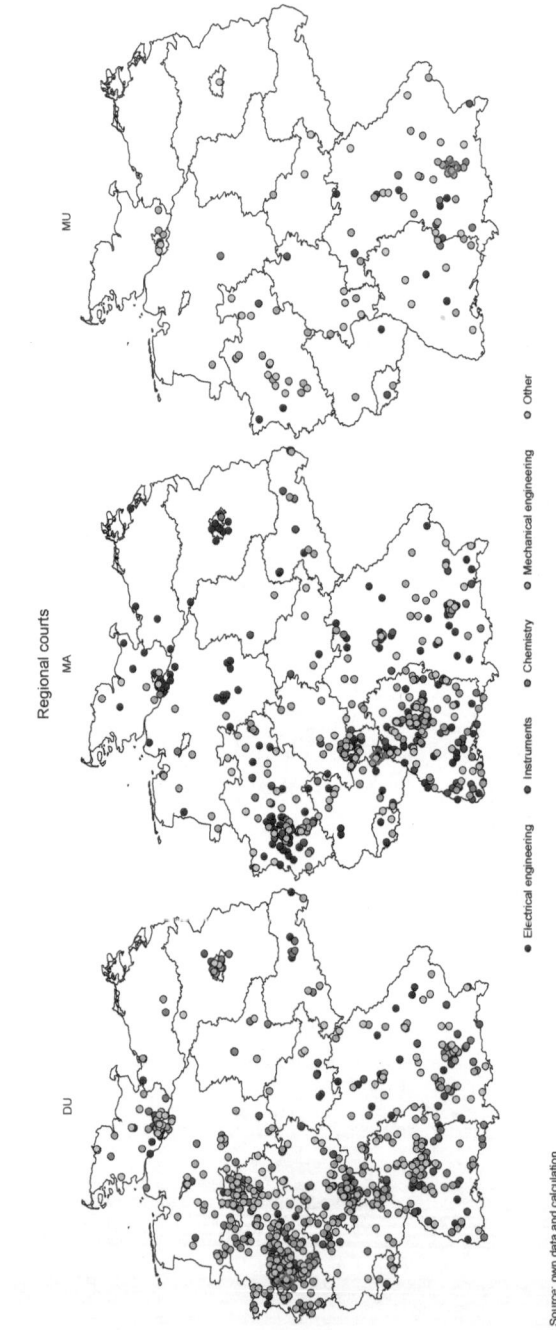

Figure 2.10: Spatial distribution of alleged infringers by court and technology main area

Notes: The sample consists of all patent infringement proceedings based on patents in force. The unit of observation is at the case level. Only alleged infringers with residence in Germany represented. In case of multiple alleged infringers, the one closest to court is chosen. Diamond symbolizes place of regional court.

Figure 2.11: Ex ante predicted length of infringement proceeding with judgment by court and technology main area (densities)

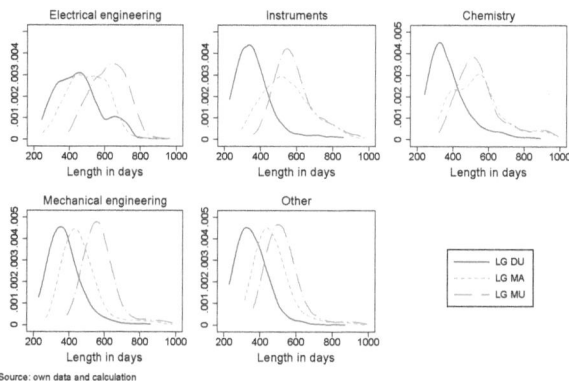

Source: own data and calculation

Notes: The sample consists of all patent infringement proceedings based on patents in force. The unit of observation is at the case level. Truncated at 1,000 days.

Figure 2.12: Ex ante predicted length of infringement proceeding with judgment by court and sample (densities)

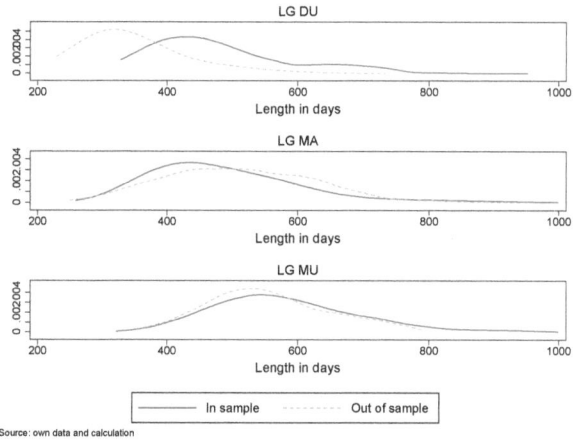

Source: own data and calculation

Notes: The sample consists of all patent infringement proceedings based on patents in force. The unit of observation is at the case level. In sample contains all proceedings actually heard by regional court. Out of sample contains all proceedings heard by another regional court. Truncated at 1,000 days.

2.8.2 Tables

Table 2.11: Statistics of litigation value by court and technology main area

Regional court	Litigation value (in thousand €)				
	Mean	Median	Std. dev.	Min	Max
Electrical engineering					
LG DU	1,809.80	1,500.00	1,768.64	2	15,000
LG MA	297.44	50.00	696.70	0	9,000
LG MU	242.15	85.00	531.76	5	3,500
Instruments					
LG DU	1,012.78	500.00	1,917.48	4	18,000
LG MA	760.85	450.00	1,171.50	12	5,000
LG MU	373.77	250.00	394.21	39	1,500
Chemistry					
LG DU	1,288.44	800.00	2,339.10	38	30,000
LG MA	899.68	500.00	2,392.97	10	16,500
LG MU	505.83	275.00	517.00	50	2,000
Mechanical engineering					
LG DU	684.35	500.00	969.71	1	10,000
LG MA	491.06	500.00	503.94	7	5,000
LG MU	403.07	275.00	443.40	5	2,000
Other					
LG DU	606.16	300.00	978.90	2	10,000
LG MA	317.40	250.00	296.64	2	1,534
LG MU	270.70	180.00	309.89	20	2,000
All	859.12	500.00	1,485.07	1	30,000

Notes: The sample consists of all patent infringement proceedings based on patents in force. The unit of observation is at the case level.

Table 2.12: Residence countries of litigants by court

Country	Regional court						Total	
	LG DU		LG MA		LG MU			
	N	%	N	%	N	%	N	%
Patent holders								
European								
AT	41	2.4%	14	2.0%	3	1.6%	58	2.2%
BE	18	1.0%	2	0.3%	0	0.0%	20	0.8%
CH	84	4.9%	26	3.8%	7	3.7%	117	4.5%
DE	812	47.2%	279	40.3%	117	62.2%	1208	46.5%
DK	32	1.9%	2	0.3%	1	0.5%	35	1.3%
FR	73	4.2%	16	2.3%	11	5.9%	100	3.8%
IE	11	0.6%	4	0.6%	1	0.5%	16	0.6%
IT	53	3.1%	250	36.1%	23	12.2%	326	12.5%
NL	93	5.4%	31	4.5%	0	0.0%	124	4.8%
NO	12	0.7%	0	0.0%	0	0.0%	12	0.5%
SE	29	1.7%	11	1.6%	4	2.1%	44	1.7%
UK	51	3.0%	4	0.6%	5	2.7%	60	2.3%
Other	21	1.2%	10	1.4%	0	0.0%	31	1.2%
Non-European								
IL	23	1.3%	3	0.4%	0	0.0%	26	1.0%
JP	170	9.9%	5	0.7%	8	4.3%	183	7.0%
US	174	10.1%	32	4.6%	4	2.1%	210	8.1%
Other	22	1.0%	3	0.4%	4	2.1%	29	1.1%
Total	1,719	100.0%	692	100.0%	188	100.0%	2,599	100.0%
Alleged infringers								
European								
AT	27	1.6%	9	1.3%	2	1.1%	38	1.5%
BE	19	1.1%	0	0.0%	0	0.0%	19	0.7%
CH	17	1.0%	6	0.9%	2	1.1%	25	1.0%
CZ	14	0.8%	0	0.0%	1	0.5%	15	0.6%
DE	1242	72.3%	563	81.4%	125	66.5%	1930	74.3%
DK	12	0.7%	0	0.0%	0	0.0%	12	0.5%
ES	15	0.9%	7	1.0%	12	6.4%	34	1.3%
FR	54	3.1%	5	0.7%	3	1.6%	62	2.4%
IT	63	3.7%	15	2.2%	9	4.8%	87	3.3%
NL	31	1.8%	9	1.3%	2	1.1%	42	1.6%
PL	21	1.2%	3	0.4%	0	0.0%	24	0.9%
SE	10	0.6%	0	0.0%	1	0.5%	11	0.4%
TR	6	0.3%	4	0.6%	1	0.5%	11	0.4%
UK	48	2.8%	2	0.3%	1	0.5%	51	2.0%
Other	23	1.3%	3	0.4%	1	0.5%	28	1.1%
Non-European								
CN	52	3.0%	18	2.6%	9	4.8%	79	3.0%
HK	10	0.6%	16	2.3%	1	0.5%	27	1.0%
KR	11	0.6%	7	1.0%	1	0.5%	19	0.7%
TW	18	1.0%	17	2.5%	5	2.7%	40	1.5%
US	8	0.5%	2	0.3%	2	1.1%	12	0.5%
Other	18	1.0%	6	0.9%	9	4.8%	33	1.3%
Total	1,719	100.0%	692	100.0%	188	100.0%	2,599	100.0%

Notes: The sample consists of all patent infringement proceedings based on patents in force. The unit of observation is at the case level. In case of multiple patent holders or alleged infringers, the one closest to court is chosen. Category Other refers to countries with N < 10 in total.

Table 2.13: Median lengths of proceedings by year

Year	Infringement (DE+EP) Length	Authority	Length	Authority	Length	Authority	Opposition (DE) Length	Authority	Opposition (EP) Length	Authority	Revocation (DE+EP) Length	Authority
2000	-	LG DU	-	LG MA	-	LG MU	47.80	DPMA	39.80	EPO	15.27	BPatG
2001	-	LG DU	-	LG MA	-	LG MU	44.80	DPMA	40.63	EPO	16.45	BPatG
2002	-	LG DU	-	LG MA	-	LG MU	20.77	BPatG	42.22	EPO	16.63	BPatG
2003	11.52	LG DU	9.03	LG MA	10.52	LG MU	29.27	BPatG	41.55	EPO	17.70	BPatG
2004	12.77	LG DU	8.57	LG MA	10.52	LG MU	37.85	BPatG	42.87	EPO	16.50	BPatG
2005	12.50	LG DU	10.12	LG MA	10.52	LG MU	44.43	BPatG	43.42	EPO	21.10	BPatG
2006	13.62	LG DU	8.55	LG MA	10.52	LG MU	49.83	BPatG	39.47	EPO	23.37	BPatG
2007	12.30	LG DU	9.28	LG MA	10.52	LG MU	27.30	DPMA	36.87	EPO	19.03	BPatG
2008	12.60	LG DU	10.03	LG MA	10.52	LG MU	27.30*	DPMA	36.87*	EPO	20.38	BPatG
2009	-	LG DU	-	LG MA	-	LG MU	27.30*	DPMA	36.87*	EPO	20.22	BPatG

* Value from prior year used to avoid truncation.

Notes: Length represents the median length in months of all respective proceedings. Year refers to year of filing. I exclude infringement and revocation proceedings that did not end by judgment from the calculations. Information on infringement: own data and calculations. Information on DPMA oppositions: PATSTAT (cf. Section 2.4.1). Information on EPO oppositions: PATSTAT (cf. Section 2.4.1). Information on BPatG oppositions: data received directly from BPatG. Information on revocations: data introduced and used in Cremers *et al.* (2014).

Table 2.14: Probit model results: incidence of expert opinion (interaction effects)

	(E1) Expert opinion	(E2) Expert opinion	(E3) Expert opinion	(E4) Expert opinion
Court effects				
LG Mannheim (d)	0.136***	0.113***	0.342***	
	(0.031)	(0.026)	(0.094)	
LG Munich (d)	0.243***	0.238***	0.652***	
	(0.050)	(0.049)	(0.117)	
Judge effects				
Judge 1 (LG MA) (d)				0.514
				(0.348)
Judge 2 (LG DU) (d)				−0.079
				(0.061)
Judge 2 (LG MU) (d)				0.240
				(0.190)
Judge 2 (LG MA) (d)				−0.037
				(0.078)
Judge 3 (LG MA) (d)				0.166
				(0.154)
Judge 3 (LG DU) (d)				−0.170***
				(0.014)
Technology effects				
Electrical engineering (d)			0.067	0.030
			(0.044)	(0.054)
Instruments (d)			0.057	−0.056
			(0.060)	(0.052)
Chemistry (d)			0.020	−0.037
			(0.048)	(0.048)
Mechanical engineering (d)			0.025	−0.014
			(0.036)	(0.045)
Court x Technology effects				
LG MA x Electrical engineering			−0.098**	
			(0.031)	
LG MA x Instruments (d)			−0.006	
			(0.078)	
LG MA x Chemistry (d)			0.111	
			(0.119)	
LG MA x Mechanical engineering (d)			−0.035	
			(0.051)	
LG MU x Electrical engineering (d)			0.006	
			(0.081)	
LG MU x Instruments (d)			0.020	
			(0.131)	
LG MU x Chemistry (d)			0.027	
			(0.108)	
LG MU x Mechanical engineering (d)			0.005	
			(0.079)	
Judge x Technology effects				
Judge 1 (LG MU) x Electrical engineering (d)				0.059
				(0.124)
Judge 1 (LG MU) x Instruments (d)				0.092
				(0.236)
Judge 1 (LG MU) x Chemistry (d)				0.085
				(0.165)
Judge 1 (LG MU) x Mechanical engineering (d)				0.056
				(0.130)
Judge 1 (LG MA) x Electrical engineering (d)				−0.029
				(0.103)
Judge 1 (LG MA) x Instruments (d)				0.211
				(0.293)
Judge 1 (LG MA) x Mechanical engineering (d)				0.062
				(0.192)
Judge 2 (LG DU) x Electrical engineering (d)				−0.035
				(0.062)
Judge 2 (LG DU) x Instruments (d)				0.255
				(0.164)
Judge 2 (LG DU) x Chemistry (d)				0.134
				(0.129)

Continued on next page

Table 2.14 – continued from previous page

	(E1) Expert opinion	(E2) Expert opinion	(E3) Expert opinion	(E4) Expert opinion
Judge 2 (LG DU) x Mechanical engineering (d)				0.097
				(0.093)
Judge 2 (LG MU) x Electrical engineering (d)				−0.021
				(0.110)
Judge 2 (LG MU) x Instruments (d)				0.212
				(0.248)
Judge 2 (LG MU) x Chemistry (d)				0.080
				(0.184)
Judge 2 (LG MU) x Mechanical engineering (d)				0.040
				(0.136)
Judge 2 (LG MA) x Electrical engineering (d)				−0.099***
				(0.027)
Judge 2 (LG MA) x Instruments (d)				0.022
				(0.115)
Judge 2 (LG MA) x Chemistry (d)				−0.004
				(0.108)
Judge 2 (LG MA) x Mechanical engineering (d)				−0.015
				(0.068)
Judge 3 (LG MA) x Electrical engineering (d)				−0.058
				(0.058)
Judge 3 (LG MA) x Instruments (d)				0.295
				(0.240)
Judge 3 (LG MA) x Chemistry (d)				0.513*
				(0.247)
Judge 3 (LG MA) x Mechanical engineering (d)				−0.019
				(0.102)
Judge 3 (LG DU) x Electrical engineering (d)				0.892***
				(0.009)
Judge 3 (LG DU) x Chemistry (d)				0.897***
				(0.008)
Other	Yes	Yes	Yes	Yes
Pseudo R^2	0.047	0.062	0.080	0.103
Observations	1,690	1,690	1,690	1,645

Marginal effects; Standard errors in parentheses
(d) for discrete change of dummy variable from 0 to 1
* $p < 0.05$, ** $p < 0.01$, *** $p < 0.001$

Notes: The sample consists of all patent infringement proceedings based on patents in force that did not end in settlement prior to the call for an expert opinion ($\bar{t}_{jt}^{ord} - 50$ days). Proceedings with judgment by consent/default decree excluded. The unit of observation is at the case level. 45 observations in regression (E4) dropped due to perfectly predicted failure. Wald test in regression (E4) jointly significant for Judge x Technology effects with one constraint dropped. Base line regional court: LG Düsseldorf (DU). Base line technology area: Other. Others: nonpatent literature backward citations (ratio), no. of claims, IPC count, age of patent, grant lag (difference from mean in days), prior infringement decision on patent (d), litigation value (in thousand €, log), multijurisdictional litigation (d), tenure as judge (in years), prior exposure to technology, and year effects. Standard errors clustered by patent.

Table 2.15: Discount factor by technology area

	Technology area				
	Chemistry	Electrical engineering	Instruments	Mechanical engineering	Other fields
Discount factor δ	0.04	0.15	0.10	0.15	0.10

2.8.3   Definition of Product Market Proximity

For corporate litigants I obtained the assigned NACE (Nomenclature of Economic Activities) Rev. 2 industry codes from company databases.[46] The official European statistical classification of economic activities since 2008, NACE Rev. 2 industry codes are 4-digit codes grouped into increasingly specified industry categories with each decimal place.[47] To quantify the product market proximity of opposing litigants, I compare their industry codes and define proximity through the match of decimals of the 4-digit code.[48]

Let a plaintiff's industry code be X and a defendant's industry be Y, with

$$X = \sum_{i=0}^{3} a_i \cdot 10^i \quad \text{and} \quad Y = \sum_{i=0}^{3} b_i \cdot 10^i.$$

The product market proximity α is then expressed as

$$\alpha = \frac{\gamma}{4} \quad \text{with} \quad \gamma = \begin{cases} 1 & \text{if } a_3 = b_3, \forall i \in \{0,1,2\} : a_i \neq b_i \\ 2 & \text{if } a_3 = b_3, a_2 = b_2, \forall i \in \{0,1\} : a_i \neq b_i \\ 3 & \text{if } \forall i \in \{1,2,3\} : a_i = b_i, a_0 \neq b_0 \\ 4 & \text{if } X = Y \\ 0 & \text{if } \forall i \in \{0,1,2,3\} : a_i \neq b_i. \end{cases}$$

I determine the product market proximity between a nonproducing entity (NPE), which by definition is not active in the product market, and any litigant as $\alpha = 0$. Litigants are classified as NPEs using the methodology found in Helmers *et al.* (2014). Litigants, especially large diversified firms, may have several assigned industry codes. In these cases, I define the product market proximity as the highest α of all industry code combinations. Likewise, infringement proceedings may have multiple litigants on both the plaintiff and defendant sides. In these cases, I define the product market proximity as the highest α of all combinations of opposing litigants.

---

[46]http://ec.europa.eu/eurostat/ramon/nomenclatures/index.cfm?TargetUrl=LST_NOM_DTL&StrNom=CL_NACE2&StrLanguageCode=EN [accessed: 22 July 2015].

[47]Where only industry codes of a different classification system were available, I transformed those into NACE Rev. 2 industry codes pursuant to official correspondence tables.

[48]This measure is highly related to yet distinct from the one used by Bloom *et al.* (2013), who capture product market proximity with industry codes weighted by their share of sales. Unfortunately, information on corporate sales at IPC level is not available in Orbis.

2.8.4   Operationalization of Opportunity Costs

Opportunity costs emerge from the time required until the plaintiff is able to legally enforce his patent. Every patent litigation dispute i is defined by a litigation value L. This litigation value is determined by the value of the patent, the remaining time of patent protection, and the scope of the infringement. The observed litigation value L_i thus equals the value of all rents the patent holder could receive during the time from the start of the dispute until the patent's expiration, T, discounted by the technology-specific factor $\delta(\delta \in [0,1])$[49]:

$$L_i = \int_0^T M_i \, (1-\delta_i)^t dt. \tag{2.8.1}$$

For the sake of simplicity I assume $B = 0$ and solve for M:

$$M_i = \frac{L_i \ln(1-\delta_i)}{(1-\delta_i)^T - 1} \tag{2.8.2}$$

The opportunity costs of dispute i at court j can then be operationalized by plugging equation (2.8.2) into equation (2.2.3) from Section 2.2.1:

$$OPC_{ij} = \alpha_i \int_0^{l_{ij}} M_i \, (1-\delta_i)^t dt$$

$$= \alpha_i \int_0^{l_{ij}} \frac{L_i \ln(1-\delta_i)}{(1-\delta_i)^T - 1} \, (1-\delta_i)^t dt$$

$$= \alpha_i \frac{L_i \left((1-\delta_i)^{l_{ij}} - 1\right)}{(1-\delta_i)^T - 1}, \tag{2.8.3}$$

where the factor $\alpha(\alpha \in [0,1])$ represents the product market proximity between the litigants.

---

[49]Future returns from a patent are commonly discounted due to the likelihood of technological obsolescence or the possibility to invent around the patent. I broadly follow the technology-specific approximate decay rates estimated in Schankerman (1998) (see also Table 2.15).

Chapter 3

Invalid but Infringed? An Analysis of Germany's Bifurcated Patent Litigation System

3.1 Introduction

Patents are probabilistic property rights: there exists inherent uncertainty regarding a patent's validity and scope (Lemley and Shapiro, 2005). Although patents are granted by patent offices only after substantive examination, there is no guarantee that a granted patent is in fact valid.[50] This uncertainty about patent validity has important effects on the functioning of patent systems and legal patent enforcement in particular.

In most patent systems, such as in the U.S. or U.K., infringement and invalidity of a patent are decided simultaneously, where infringement is only possible if the patent is upheld in the same proceeding.[51] In so-called bifurcated patent litigation systems, such as in Germany, China, or Japan, however, separate courts decide on infringement and validity independently. Because challenging a patent's validity requires an additional, separate court proceeding, alleged infringers are less likely to do so. This means that in a bifurcated system, there is a higher likelihood of seeing an invalid patent enforced. Even if a validity challenge is filed, in practice, the decision on infringement is often rendered and enforced before validity has been

---

[49]This chapter is joint work with Katrin Cremers, Dietmar Harhoff, Christian Helmers, and Yassine Lefouili. The author of this dissertation was primarily responsible for the institutional part (Section 3.2) and the empirical analysis (Section 3.4 and Section 3.5).

[50]Mann and Underweiser (2012), for example, show that since 2003 the U.S. Federal Circuit has held nearly 60% of patents considered invalid.

[51]In the U.S., courts decide on both infringement and validity simultaneously. However, the Inter Partes Review (IPR), which was introduced by the America Invents Act (AIA) in September 2012 as a way of challenging validity administratively at the U.S. Patent and Trademark Office post-grant, has *de facto* introduced bifurcation into the U.S. system (Chien and Helmers, 2015).

determined under the presumption that granted patents are indeed valid. This means that a bifurcated enforcement system prioritizes resolving uncertainty regarding infringement. This has advantages; perhaps most importantly it promotes prompt decisions on patent infringement. But because patents are probabilistic rights, this can lead to situations in which a patent is held infringed that is subsequently invalidated. Our objective is primarily to empirically quantify the extent to which bifurcation deters validity challenges and creates such 'invalid but infringed' decisions. Also, we explore the potential implications of the uncertainty that they create by looking at changes in preemptive validity challenges in the form of post-grant oppositions.

Using detailed case level data from German courts where infringement and validity are separated into independent proceedings, we show that in practice the decision on infringement is often rendered and enforced, before validity has been determined under the presumption that granted patents are indeed valid. We show that this leads to situations in which a patent is held infringed that is subsequently invalidated. Our data on infringement and invalidity proceedings in Germany for 2000 to 2008 reveals that 12% of infringement cases with parallel invalidity proceedings (42% if we focus on cases with decision in both venues) produce such 'invalid but infringed' decisions.[52] Our analysis also shows that the length of this *injunction gap* is substantial. In cases where validity was challenged in court, the infringement decision was on average enforceable for more than a year before the patent was first invalidated in first instance. These results show that bifurcation offers scope for patent holders to temporarily enforce invalid patents.

We also show that bifurcation reduces the likelihood that an alleged infringer challenges a patent's validity in the first place. Because challenging the validity of a patent requires a separate action at a different court, the alleged infringer may refrain from doing so despite the potential invalidity of a patent. We find evidence that smaller and foreign firms in particular are less likely to file a validity challenge at the German Federal Patent Court (BPatG) when an infringement action is brought against them. The effect is robust to controlling for various time varying and invariant patent, litigant, and case level characteristics. This suggests that more resource-constrained firms are less likely to challenge a patent's validity. The implications of this *screening effect* are twofold: the share of cases where an infringed patent is invalidated is downwardly biased, and the strong presumption of validity built into the bifurcated litigation system becomes self-reinforcing.

We use a simple model to demonstrate that the possibility of patent enforcement with little delay and the lower likelihood of facing a counterclaim for invalidity favor the patent holder

---

[52]Table 3.10 in the Appendix provides a list of 'invalid but infringed' decisions drawn from our data.

suing for infringement. If patents serve as an incentive mechanism to encourage investments in innovation, strong rights to enforce a patent against alleged infringers may be even socially desirable. However, the model shows that in a bifurcated system – provided there is some uncertainty with regard to the validity of a patent – the patent holder's incentives are distorted. These distortions originate in (a) the possibility to enforce an invalid patent during the injunction gap and (b) the lower likelihood of facing a validity challenge for a potentially invalid patent.

Bifurcation can also create uncertainty. The main argument is that the likelihood to be found to have infringed an invalid patent is higher in a bifurcated system than in a system where validity and infringement are assessed in the same proceedings. While the time lag between the decisions on infringement and validity in itself creates uncertainty for the litigants, more fundamentally, the increased likelihood of being found to have infringed an invalid patent creates additional uncertainty for firms when navigating the patent landscape. This could potentially have important effects on firms' innovation and patenting activities. For instance, firms that have fallen into the injunction gap might alter their assessment of the likelihood of facing an injunction despite the invalidity of allegedly infringed patents. Such changes in perceptions remain unobservable to us.[53] However, we can test whether they manifest themselves in changes in opposition behavior of alleged infringers, i.e., we can test whether firms subject to a divergent decision oppose more patents immediately following this experience. Our results show that alleged infringers are indeed more likely to file oppositions after they have experienced a divergent decision. We interpret this as evidence that firms attempt to preempt similar situations in the future by eliminating potentially threatening patents early on. This finding is consistent with the fact that German firms are overall responsible for a disproportionately large share of oppositions at the EPO, and suggests that this partly reflects the uncertainty created by the bifurcated litigation system.

Our research contributes to the existing literature on the design and functioning of patent litigation systems by offering first quantitative evidence on the implications of bifurcation. This is not only of direct relevance to Germany, where the large majority of patent cases in Europe are litigated,[54] but also plays an important role in the current heated discussion about the design of the Unified Patent Court (UPC) in Europe. For example, a group of large firms across

---

[53]There is, however, anecdotal evidence. For example, in April 2012, Microsoft announced the relocation of its European logistics center from Germany to the Netherlands citing the threat of a possible injunction due to the alleged infringement of a Motorola patent (various news sources, including Reuters, the Wall Street Journal, and the Financial Times, 2 April 2012.) Microsoft considered the risk of facing an injunction to be higher in Germany than in the Netherlands, presumably due to the bifurcated litigation system.

[54]Cremers *et al.* (2013) show that depending on how cases are counted (e.g., counting infringement and revocation cases as separate cases or not) the total number of patent cases in Germany is between 12 and 29 times larger than in the UK.

various industries, including Adidas, Apple, Deutsche Post DHL, Google, and Samsung,[55] issued a joint statement in February 2014 voicing concerns that: "[...] the potential exists for a court to order an injunction prohibiting the importation and sale of goods even though the patent may ultimately be found invalid. This result unduly reduces competition, can increase the cost of products in the market and reduce product choices, all negatively impacting consumers."

From a broader perspective, our evidence underscores the probabilistic nature of patents. We show that granted patents which a court presumes valid when deciding on infringement often turn out to be invalid under closer scrutiny. Patents involved in court disputes are only the tip of the patent iceberg and clearly a non-random selection. Regardless, our evidence supports the general view that legal rights in the form of patents are inherently associated with enormous uncertainty. We also show that bifurcation compounds the undersupply of validity challenges in court that has been shown to exist in non-bifurcated systems (Farrell and Merges, 2004). This means that the strong presumption of validity of probabilistic patents distorts incentives to the rights holder's advantage. This offers empirical evidence directly relevant for the long-standing, largely theoretical debate on the design of patent litigation systems (Aoki and Hu, 1999; Ayres and Klemperer, 1999; Boyce and Hollis, 2007).[56]

The remainder of this study is organized as follows: the next section provides a detailed description of the German patent litigation system with particular focus on the interplay between infringement and invalidity proceedings. Section 3.3 discusses the benefits of a system with a strong, built-in presumption of validity and presents a model to discuss potential legal discrepancies and how they may translate into increased uncertainty. Section 3.4 describes the data used in our analysis. Section 3.5 presents our empirical findings, and Section 3.6 offers some concluding thoughts and suggestions for further research.

3.2 Germany's Bifurcated Patent Litigation System

This section explains the design of the German bifurcated patent litigation system, with a focus on the legal framework that can lead to divergent decisions in infringement and invalidity proceedings.

---

[55]The complete list is: Adidas, AFDEL, Apple, ARM, BlackBerry, Broadcom, Bull, Cisco Systems, Dell, Deutsche Post DHL, ESIA, Google, HP, Huawei, Microsoft, Samsung, SFIB, Telecom Italia, and Vodafone.

[56]See also Weatherall and Webster (2014) for a review of the literature on legal patent enforcement.

3.2.1 Court Structure

Regional courts (*Landgerichte* – LGs) have jurisdiction over patent infringement.[57] There are twelve regional courts that serve as first instance courts in infringement proceedings.[58] A panel of three legally trained judges decide on infringement. Decisions by the regional courts can be appealed before a higher regional court (*Oberlandesgericht* – OLG). In exceptional cases, a further appeal can be brought before the Patent Division of the German Federal Court of Justice (*Bundesgerichtshof* – BGH) in third instance.

A patent's validity is challenged either through opposition filed at the patent office which granted the patent right (European Patent Office – EPO – for *EP* patents or *Deutsches Patent-und Markenamt* – DPMA – for *DE* patents) or through a revocation action filed at the German Federal Patent Court (*Bundespatentgericht* – BPatG).[59] As a specialized court, the BPatG deploys judges with both legal and technical training.[60] Appeals against decisions by the BPatG are directly brought before the Patent Division of the Federal Court of Justice (*Bundesgerichtshof* – BGH) that reviews infringement proceedings. The structure of the German court system is summarized in Figure 3.6 in the Appendix.

Infringement

The patent holder filing an infringement action can seek different forms of legal relief; for example, a cease and desist order to halt the infringing act, the recall and destruction of infringing goods, rendering of account to identify distribution channels and calculate damages, or damages for losses suffered. The patent holder can also request a preliminary injunction against the alleged infringer. However in practice, preliminary injunctions are rare because they require clear-cut evidence regarding the infringing act, the validity of the patent, and urgency (cf. Kühnen, 2012).[61]

The main oral hearing takes place roughly six to twelve months after the action was filed. Main oral hearings rarely exceed a day in length and often last for only a few hours. In case

---

[57] Infringement claims must be based on a patent granted by the DPMA (*DE*) or the EPO with effect for Germany (*EP*).

[58] These are the regional courts in Berlin, Braunschweig, Düsseldorf, Erfurt, Frankfurt, Hamburg, Leipzig, Magdeburg, Mannheim, Munich, Nuremberg-Furth, and Saarbrücken. Each regional court has at least one chamber primarily designated to patent cases.

[59] The responsibilities of the BPatG are twofold. It serves as the appeals court for decisions of the DPMA concerning *DE* patent applications, and it hears revocation actions for *DE* and *EP* (with effect for Germany) patents.

[60] The panel consists of five judges: three technically trained judges and two legally trained judges.

[61] An injunction might be granted, for example, if the suspected infringer is about to start selling a product that clearly infringes a patent that is most likely valid and where selling the infringing product would result in substantial losses for the patent holder. Although there has been a recent increase in the number of preliminary injunctions (Müller-Stoy and Wahl, 2008), they are still a relatively rare occurrence in patent litigation (Böhler, 2011).

of a parallel validity challenge, the judges may grant the request to stay the proceeding until a decision on the patent's validity is available (see Section 3.2.2 below). If the infringement action is not stayed, the judges usually render their decision four to eight weeks after the main oral hearing. Alternatively, the litigants may settle at any time during the proceeding. The prevailing litigant can demand reimbursement of his legal costs from the losing party.

In the proceeding, the defendant may dispute the infringement allegations, but a possible invalidity of the patent does not constitute an admissible defense. The alleged infringer must challenge the patent's validity through a separate opposition or revocation action.

Invalidity

The alleged infringer may challenge a granted patent through opposition or, subsequently, a revocation action. An opposition to an *EP* (*DE*) patent can be filed at the EPO (DPMA) within the first 9 months (3 months) after grant of the patent. After this period, the alleged infringer may still join an already pending opposition proceeding. It is noteworthy that the EPO and DPMA may continue the proceeding *ex officio* and decide on validity even if the opponent withdraws the opposition. If invalidated, the patent is deemed void, counting from its grant date.[62] Each litigant usually bears his own costs in an opposition proceeding.

After the end of the opposition phase, or – in case of an opposition – after the end of the opposition proceeding, validity can be challenged only through a revocation action at the BPatG. Although a revocation action can be filed by any person or legal entity, almost all revocation actions are filed in response to infringement actions.[63]

Unlike in opposition proceedings, the plaintiff has full discretion to withdraw his action at any time. As with oppositions, if the BPatG invalidates a patent, it is deemed void from its grant date. The prevailing party in the revocation proceeding can demand reimbursement of his legal costs from the losing party.

3.2.2   Interaction of Infringement and Invalidity Proceeding

If a patent is invalidated, any pending infringement proceedings based on the patent will be dismissed.[64] This still allows for situations where decisions on infringement can be (preliminarily) enforced based on an invalid patent if infringement is decided before invalidity. The

---

[62]Note that for *EP* patents, the decision has effect in all EPC countries where the opposed patent is validated.

[63]von Hees and Braitmayer (2010) estimate that this is the case for 90% of all revocation actions.

[64]If the patent is only partly invalid, the subject matter in pending infringement proceedings has to be reconsidered on the basis of the amended patent. However, if the infringement proceeding is no longer pending, the alleged infringer has to demand a reconsideration of the case on the basis of the amended patent by filing a separate restitution action.

occurrence of such divergent decisions crucially depends on (a) the timing and (b) the duration of infringement and invalidity proceedings:

(a) As a defensive reaction to an infringement action, validity challenges are usually filed after the corresponding infringement proceedings. This is often due to the time required to prepare the case, in particular the search for prior art that can be used to challenge the patent's validity (Kühnen, 2013).[65]

(b) Revocation proceedings take significantly longer than infringement proceedings in first instance (see Figure 3.7 in the Appendix), thus increasing the temporal gap between the decisions. Taking into account a possible appeal, litigants have to expect a maximum of five to seven years until a final judgment on a revocation action. Opposition proceedings also take significantly longer than infringement proceedings. The litigants may request acceleration of the proceeding, but an opposition still takes 30 to 40 months on average.[66]

In combination, (a) and (b) mean that decisions on invalidity follow infringement decisions with a considerable lag.

Figure 3.1: Timing of infringement and invalidity proceedings in bifurcated and non-bifurcated systems (own illustration)

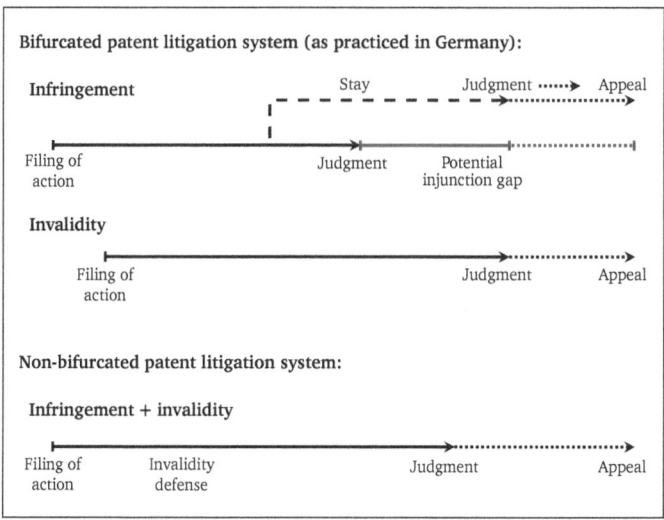

---

[65]We find that more than 55% of parallel invalidity proceedings are initiated at least four months after the corresponding infringement proceeding.
[66]Harhoff and Reitzig (2004) report a median length of opposition proceedings at the EPO of about four years (including appeal).

The alleged infringer can request to stay the infringement proceeding until a decision on validity is available (see Figure 3.1). In their decision to grant a stay, the infringement court judges attempt to strike a balance between the inherent conflict of interest between the litigants.[67] On the one hand, the alleged infringer may suffer irrecoverable damage if she is found to infringe an invalid patent. On the other hand, the delay to the judgment on infringement can be considerable if a stay is granted. In this case, the patent holder would be prevented from timely enforcement.

In practice, infringement courts rely by case law on a strong presumption of validity. That is, infringement proceedings are only stayed if there is an overwhelmingly large probability that the patent will not be upheld in its current form. So, even though the judges at the infringement courts do not consider the validity of the patent in their judgment on infringement, they have to form an opinion on the likelihood of invalidity to decide on a stay (Fock and Bartenbach, 2010). This poses a considerable challenge, as infringement court judges are rarely technically trained, and limited resources restrict a thorough investigation of the patent's validity. Usually, the corresponding validity challenges are not yet at a stage where they could provide guidance on the likelihood of invalidity.[68] Infringement court judges are therefore forced to stay at their own discretion.

3.3   Effects of Bifurcation

Proponents of bifurcation argue that a centralized jurisdiction for patent validity offers the advantage of specialization. The court in charge of validity cases can train and deploy technical judges and accumulate experience specifically in the assessment of patent validity. This should result in coherent and well-founded claim construction and thus increase legal certainty regarding the validity of patents. Another argument in favor of bifurcation is the screening effect. Separate patent invalidity proceedings increase the costs and risks for the alleged infringer. If the alleged infringer expects a patent to be upheld, she will refrain from a validity challenge as a defense to avoid further expenses. Accordingly, bifurcated patent litigation systems deter validity challenges with relatively low chances of success. Perhaps the most important argument is that a strong presumption of validity, which puts considerable faith in the pre-grant

---

[67]For example, judges take the expected length of a stay into consideration when deciding whether to stay infringement proceedings (Kaess, 2009). A stay is usually not granted if the prior art forwarded has already been considered in the patent examination or any prior invalidity proceedings. Further factors taken into consideration can be found in Harguth and Carlson (2011) and Kühnen (2012).

[68]With the Patent Law Revision Act introduced in 2009, the BPatG is now supposed to provide an interim assessment of the patent's validity as soon as possible. The infringement court, however, is not bound to the assessment. Note that our data predate this revision of the law.

examination of patent offices, leads to little delay in judgment on infringement allegations since validity is not assessed simultaneously. In combination, fast decisions on infringement, the deterrence of futile validity challenges, and the specialized institutions that decide on the technical question of validity promise to offer efficient enforcement of patents (see e.g., Hilty and Lamping, 2011).

Conversely, opponents of bifurcation argue that a system separating infringement and invalidity proceedings is prone to legal discrepancies. We discuss these and show their implications below.

3.3.1 Divergent Decisions and Screening Effect

Even though infringement and invalidity proceedings are heard and decided by different courts, the decision on the patent's validity has consequences for the decision on infringement, provided infringement was found. Once a patent is invalidated, this decision erodes the legal basis for any claims for infringement. At the same time, because decisions on infringement are usually rendered faster than decisions on validity, a court may find infringement even though the patent is later invalidated. In fact, if infringement is found in first instance, any injunction resulting from this decision is enforceable regardless of appeal or any pending validity challenge. This means the greater the temporal gap between infringement and validity decisions, the longer a patent can be wrongfully enforced. Even if the patent is invalidated in first instance, the patent holder can continue to enforce the patent as long as the decision does not become binding. The injunction gap may, therefore, extend beyond the first instance invalidity decision. This again creates strong incentives to appeal the infringement decision while awaiting the outcome of the validity challenge. The result is considerable legal uncertainty about the outcome of the infringement dispute, potential delays in enforcement, increased litigation costs, and the possibility of an injunction gap (area A in Figure 3.2).[69]

The existence of an injunction gap has important consequences for the litigation behavior of patent holders and alleged infringers. Most importantly, an injunction gap compounds the increased incentives (relative to a non-bifurcated system) for the patent holder to sue for infringement. To see this more formally, we present a simple model that compares bifurcated

---

[69]Two measures that counteract these factors are readily cited by the proponents of bifurcation (Pitz, 2011). First, the strict separation between infringement and validity can be weakened by staying infringement proceedings until the revocation or opposition outcome becomes available. The problem with this mechanism is that the infringement court has to form an opinion on the likelihood of the validity challenge without proper assessment. Second, the alleged infringer subject to a divergent decision can obtain relief through appeal or a claim for restitution in case the patent holder has exercised an injunction on the basis of the finding for infringement (Kühnen, 2009). In this situation, the alleged infringer also has the right to demand compensation for accrued losses. Compensation may reduce the direct harm caused by an injunction, but some injunctions, such as rendering accounts to a competitor, may cause irreparable damage (cf. Ann *et al.*, 2011).

Figure 3.2: Divergent decisions

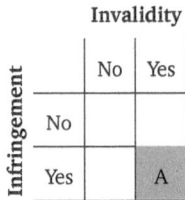

Notes: The figure shows the share of cases where infringement was either found (Yes) or not (No) and where the patent was either invalidated (Invalidity – Yes) or not (Invalidity – No). Area *A* shows the share of 'invalid but infringed' cases.

and non-bifurcated litigation systems.

Benchmark: litigation in a non-bifurcated system

We consider a patent holder (plaintiff) P and an alleged infringer (defendant) D. The patent holder can decide whether to sue the alleged infringer. If he does so, the alleged infringer can file a counterclaim to invalidate the patent without incurring any additional cost.[70] Let C_P and C_D denote the litigation costs of the patent holder and the alleged infringer respectively, θ the probability that a court finds the defendant to infringe the patent and α the probability that the patent's validity is upheld if challenged.

In this benchmark model we assume that whenever the patent's validity is challenged, the issues of infringement and validity are decided simultaneously by the same court.[71] If the court finds the patent is invalid, the alleged infringer is not affected; however, the patent holder incurs a loss L_p (e.g., loss of third party license fees). If the court finds that the patent is valid and infringed, the patent holder derives a benefit B_p which captures, for instance, damages paid by the infringer or additional market profits due to an injunction. If the court finds that the patent is valid but not infringed, we assume that neither the alleged infringer nor the patent holder are affected by this decision.[72] For now, we assume that all parameters are common knowledge.[73]

The timing of the game is as follows:

---

[70]Our results hold if we assume that there is a positive additional cost to filing a counterclaim under the reasonable assumption that this additional cost is lower under the non-bifurcated system than under the bifurcated system.

[71]It is not crucial to assume that the same court decides upon infringement and validity. What matters is that the payoffs of the parties only depend on whether the patent is infringed and valid.

[72]In Extension 3.7.3 in the Appendix we relax this assumption. We assume that if the court finds the patent to be valid but not infringed, the patent holder derives a benefit from this decision (e.g., increased deterrence of future infringements) while the defendant remains unaffected.

[73]This assumption will be relaxed in Extension 3.7.3 in the Appendix.

- **Stage 1:** The patent holder P decides whether to file an infringement action against the alleged infringer D. If he does not file for infringement, the game ends. If he does, the game proceeds to the next stage.

- **Stage 2:** The alleged infringer D decides whether to challenge the patent's validity.

- **Stage 3:** The court decides *simultaneously* on infringement and validity (provided the patent's validity was challenged).

Because challenging the patent's validity does not induce any additional costs for the alleged infringer in this benchmark model while decreasing his expected losses, the alleged infringer will always challenge the patent's validity as part of his defense. Therefore, the plaintiff will sue for infringement if and only if

$$\theta \, \alpha B_p - (1-\alpha) L_p - C_p \geq 0,$$

that is,

$$C_p \leq \theta \, \alpha B_p - (1-\alpha) L_p \equiv \tilde{C}_p^{nb}. \tag{3.3.1}$$

Litigation in a bifurcated system

In the bifurcated patent litigation system, validity and infringement decisions are rendered separately and independently (see Section 3.2 above). We assume that two different courts deal with infringement and validity, and that the court dealing with infringement hands down its judgment first. To account for the main features of a bifurcated system, we modify the benchmark setting in a number of ways.

First, we consider an alternative timing:

- **Stage 1:** The patent holder P decides whether to file an infringement action against the alleged infringer D. If he does not file for infringement, the game ends. If he does, the game proceeds to the next stage.

- **Stage 2:** The alleged infringer D decides whether to challenge the patent's validity.

- **Stage 3:** The court dealing with infringement hands down its judgment.

- **Stage 4:** The court dealing with validity hands down its judgment (provided the patent's validity was challenged).

Second, we assume that challenging the patent's validity induces additional litigation cost c_D for the alleged infringer. Third, we suppose that the alleged infringer incurs a loss l_D with $l_D \leq L_D$ if the patent is found infringed and eventually invalidated, where L_D represents the loss in case the defendant is found to infringe a valid patent. Fourth, we assume that the patent holder incurs a loss L_P if the patent is neither infringed nor valid. The patent holder receives a payoff $b_P - L_P$, where $0 \leq b_P \leq B_P$, if the patent is found infringed but eventually invalidated. For instance, the patent holder receives $b_P > 0$ if he enforces the patent during the injunction gap.[74]

In this setting, an alleged infringer will challenge the patent's validity if and only if her expected payoff from doing so $-\theta \alpha L_D - \theta (1 - \alpha) l_D - C_D - c_D$ is greater than her expected payoff from facing the infringement proceeding without challenging the patent's validity, $-\theta L_D - C_D$. Therefore, the patent's validity will be challenged if and only if

$$c_D \leq \theta (1 - \alpha)(L_D - l_D) \equiv \tilde{c}_D. \tag{3.3.2}$$

This condition shows that whenever $c_D > 0$ there are parameters for which the patent's validity is not challenged. Note that this holds even if the losses l_D incurred by the alleged infringer during the injunction gap are zero (i.e., she is able to fully recover them through compensation if the patent is invalidated). Patent validity challenges are thus less likely under a bifurcated than a non-bifurcated system.

Now we turn to the patent holder's decision to file for infringement. We need to distinguish between two cases:

- **Case 1:** $c_D \leq \tilde{c}_D$

  In this case, the patent holder expects an infringement proceeding to induce a validity challenge and therefore will file an infringement action against D if and only if

  $$\theta \alpha B_P + \theta (1 - \alpha)(b_P - L_P) - (1 - \theta)(1 - \alpha) L_P - C_P \geq 0.$$

  which can be rewritten as

  $$C_P \leq \theta \alpha B_P - (1 - \alpha) L_P + \theta (1 - \alpha) b_P = \tilde{C}_P^{nb} + \theta (1 - \alpha) b_P.$$

- **Case 2:** $c_D > \tilde{c}_D$

---

[74]Note that the non-bifurcated system can be regarded as a special case of the bifurcated system, in which the three parameters c_D, l_D and b_P are zero. Here, stages 3 and 4 collapse into a single stage.

In this case, the patent holder anticipates that there will be no validity challenge in response to his infringement action. Therefore, he files an infringement action against D if and only if

$$\theta B_P - C_P \geq 0$$

that is,

$$C_P \leq \theta B_P = \tilde{C}_P^{nb} + (1-\alpha)(\theta B_P + L_P).$$

We summarize the two cases as follows:

$$\tilde{C}_P^b = \begin{cases} \tilde{C}_P^{nb} + \theta(1-\alpha)b_P & \text{if} \quad c_D \leq \tilde{c}_D \\ \tilde{C}_P^{nb} + (1-\alpha)(\theta B_P + L_P) & \text{if} \quad c_D > \tilde{c}_D \end{cases}$$

Therefore, patent holder P files for infringement if and only if

$$C_P \leq \tilde{C}_P^b \tag{3.3.3}$$

where $\tilde{C}_P^b > \tilde{C}_P^{nb}$ holds for any $b_P > 0$ and $L_P > 0$.

The model predicts that the patent holder is more likely to file an infringement action against the alleged infringer D in a bifurcated system. There are two distinct reasons for this, depending on whether the additional costs associated with challenging the patent's validity are high or low:

(a) Low costs $c_D \leq \tilde{c}_D$: the patent holder expects the defendant to challenge the patent's validity but knows that even with an invalid patent there is a chance to obtain some benefits during the injunction gap. That is, the patent holder's incentives to enforce his patent in a bifurcated system *exceed* those in a non-bifurcated system by $\theta(1-\alpha)b_P$, which represents the pay-off from his ability to enforce an invalid patent during the injunction gap.

(b) High costs $c_D > \tilde{c}_D$: the patent holder anticipates that the defendant will not challenge the patent's validity, which in turn increases the expected benefit from filing for infringement relative to a scenario in which the patent's validity is challenged. That is, the patent holder's incentives to enforce his patent in a bifurcated system *exceed* those in a non-bifurcated system by $(1-\alpha)(\theta B_P + L_P)$, which represents the loss to the patent holder if a patent is invalidated by the court.

Hence, a patent holder's incentives to file for infringement are larger in a bifurcated than in a non-bifurcated system, because (a) bifurcation allows the enforcement of an invalid patent and (b) it deters validity challenges. This implies the bifurcated system distorts the incentives to enforce a patent to the patent owner's advantage. Finally note that since $\tilde{c}_D$ decreases in l_D, the larger the losses incurred by the defendant in the injunction gap, the more important effect (b) becomes.

Empirical predictions

Empirically, we cannot directly quantify the extent to which bifurcation distorts a patent holder's (alleged infringer's) incentives to file an infringement (invalidity) action, as this would involve a counterfactual analysis.

However, to test for the importance of (a), i.e., bifurcation allows for the enforcement of an invalid patent, we can empirically study the frequency and duration of the injunction gap, which can be seen as a proxy for both parameters l_D and b_P. Since these parameters crucially affect the magnitude of the distortions created by bifurcation, measuring the size of the injunction gap (i.e., the size of area A in Figure 3.2) provides an indication of the magnitude of the effects of (a) in a bifurcated system.

The model also offers a way of testing the presence of (b), i.e., bifurcation deters validity challenges, as it shows that the patent holder's incentives to file for infringement crucially depend on the defendant's costs in challenging the patent's validity. While we do not have direct measures of the costs involved, we can relate them to the litigant's size and location. Denote the size of the alleged infringer as s_D and denote i_D her location, which has the value of 1 if she is located in Germany and 0 otherwise. Viewing the costs of challenging a patent's validity $c_D(s_D, i_D)$ as a function of size and location, it is reasonable to assume that access to justice is less costly for domestic firms: $c_D(s_D, 1) < c_D(s_D, 0)$. Moreover, there is plenty of evidence that smaller firms are more resource-constrained (Carpenter and Petersen, 2002; Hall, 2002) and hence litigation costs may weigh heavier on them. We therefore assume that $\partial c_D / \partial s_D < 0$. Under these assumptions, our model generates two predictions that will be tested in Section 3.5. First, smaller firms are less likely to challenge the validity of a patent. Second, the probability that an alleged infringer challenges the patent's validity is higher when the alleged infringer is located in Germany.[75]

---

[75]In Section 3.7.3 in the Appendix we show that these theoretical predictions also hold for a number of extensions of the baseline model.

3.3.2   Uncertainty and Changes in Opposition Behavior

Because of its static nature, our model does not provide predictions for how firms might adapt their future behavior in the face of an injunction gap. Nonetheless, since we find that divergent decisions arise in a substantial number of cases (see Section 3.5.1), it is worth empirically studying whether those affect the alleged infringers beyond the immediate consequences of legal enforcement.

One possible effect could be a change in the alleged infringers' opposition behavior. While oppositions can be a reaction to the allegation of infringement, they are also considered common preemptive means against competitors' newly granted patents. Compared to revocation proceedings, oppositions are cheap and centralized for *EP* patents at the EPO (Mejer and van Pottelsberghe de la Potterie, 2012). However, the main rationale is that firms can curb uncertainty in the patent landscape prior to their own investments in commercializing a technology. In this way, firms can prevent future infringement allegations that may lead to unfavorable outcomes due to bifurcation.

We therefore test whether the opposition behavior of firms changes immediately after they experience an injunction gap. We have data on the entire opposition history of firms at the EPO and can check whether firms' filing activities change within a one-year window following a validity decision (i.e., once they learn that they have been subject to an injunction based on an invalid patent). To account for any general tendency to change opposition filings following litigation, we match a control group of firms that was also involved in both infringement and invalidity proceedings, but where proceedings did not end in divergent decisions. This allows us to obtain difference-in-difference estimates of any effect of the injunction gap on opposition filings. Testing for a change in the effort of alleged infringers to clear the patent landscape immediately after an injunction gap, we determine whether divergent decisions create additional uncertainty.

3.4   Data

Our data are based on patent infringement and revocation proceedings filed at German courts between 2000 and 2008 and opposition proceedings filed at the EPO and DPMA. We combine the case level information with patent and litigant data.

3.4.1   Data Sources

Regional Courts – Infringement

We collected data on infringement actions directly from the three regional courts that deal
with the majority of patent infringement cases in Germany: the Düsseldorf, Mannheim, and
Munich regional courts. We obtained detailed information on proceedings filed between 2000
and 2008. This provides us with a nine-year window, but also minimizes the number of cases
that were still pending during the data collection.

The information extracted for each case primarily concerns procedural factors, the identity
of the litigants and their legal representatives, and the patents at issue. For procedural factors,
we have data on the dates of filing, oral hearing, and judgment, as well as on the outcome, i.e.,
how the proceedings ended in each instance. We also have information on the claims made
by the plaintiff and the litigation value set by the court. Furthermore, the data include information on the names and addresses of the plaintiffs and defendants, which allowed corporate
litigants to be matched with firm level databases, including Bureau van Dijk's Orbis, Compustat and THOMSON One. In this way we gained information on firm characteristics, including
the number of employees, total assets, turnover and industry.[76] We collected information on
the litigants' legal representatives. We used this information to create a binary variable that
indicates whether a litigant was represented in court by a top law firm.[77]

With the patent application (or publication) numbers referenced in the case files, we retrieved detailed information on the litigated patents from EPO's PATSTAT.[78] PATSTAT provides
us with information on application and publication dates, IPC classes, family size, as well as
forward and backward citations. Based on the patent numbers we constructed the respective
patent families to obtain other European national and *EP* equivalents in order to identify cases
where a particular patent dispute spread across multiple national jurisdictions.[79]

Federal Patent Court – Revocation

We also have information on revocation proceedings before the BPatG and its appeal court,
the BGH. Both courts publish all decisions on validity rendered since 2000 on their websites.
We obtained information on the filing date and withdrawn actions in both instances from the
register of the German Patent and Trademark Office. This allowed us to construct the course

---

[76]The data also allow us to distinguish between natural and legal persons, such as firms, research institutions,
universities, etc.

[77]We identified top law firms in patent litigation using a ranking of leading law firms published in 2009 by the
legal professional journal *JUVE Rechtsmarkt*.

[78]We use PATSTAT version October 2012.

[79]Details on the identification of multi-jurisdictional patent disputes can be found in Cremers *et al.* (2013).

of the revocation proceedings without having to access the case files stored at the courts. That said, we could not obtain information on the party challenging the patent since the published decisions are anonymized. We link infringement and revocation proceedings based on the patents involved and not based on the litigating parties. Earlier studies (cf. Stauder, 1983) and interviews with practitioners support our assumption, that generally speaking, the alleged infringer files the revocation action as a counterclaim to an infringement allegation.

To account for revocation proceedings parallel to the infringement proceedings, but with a filing date either before or after the infringement claim, our data on revocation proceedings cover the entire 1983 to 2012 period for litigated patents.

EPO and DPMA – Opposition

The data on oppositions come from two authorities. For national *DE* patents we have information on the opposition proceeding, i.e., the opposition's filing and end dates as well as outcome, from the DPMA register. We constructed a dataset on oppositions at the EPO based on legal status information from PATSTAT covering the period 1981 to 2012. In contrast to the data from the DPMA, the data for oppositions at the EPO have information on the identity of the opponent, i.e., the party filing the opposition.

We linked the opposition data to our main dataset as follows. First, we added information on any opposition to the patents involved in an infringement proceeding to identify parallel invalidity proceedings in the form of oppositions and to construct each patent's history of validity challenges. Second, we manually identified the litigants from the infringement proceedings from all opponents of *EP* patents to capture the opposition behavior of the alleged infringers over time. We also matched the opponents with firm level data from Bureau van Dijk's Orbis.

3.4.2 Sample Description

The patent infringement cases collected at the Düsseldorf, Mannheim, and Munich regional courts cover around 80% of all patent infringement cases during the period 2000 to 2008.[80] In total, we have data for 5,121 litigation cases. We identify and exclude cases from our dataset that are not directly concerned with infringement.[81] We also drop cases involving utility models because the bifurcation principle only applies to invention patents. Furthermore, to avoid misinterpretation of case outcomes, we also remove a small number of negative declaratory

---

[80]We estimate that roughly half of the remaining 20% of cases are spread among the other nine regional courts. However, these courts are of minor importance and reputation.

[81]These represent employee invention disputes, licensing and patent transfer disputes, as well as patent arrogations and false marking claims.

actions and adjacent proceedings where the court decides only on issues regarding the enforcement of a previous judgment (e.g., rulings on costs or damages). The resulting sample contains 3,374 patent infringement cases. As some actions are filed on the basis of more than one patent, we have a total of 3,706 patent case observations.

For the period 2000 to 2008, our data show 1,822 revocation actions filed at the Federal Patent Court.[82] We also have data on all oppositions filed at the EPO between 1981 and 2012. These data cover oppositions to the granting of 68,259 *EP* patents.[83]

Figure 3.3: Incidence of infringement and parallel invalidity proceedings

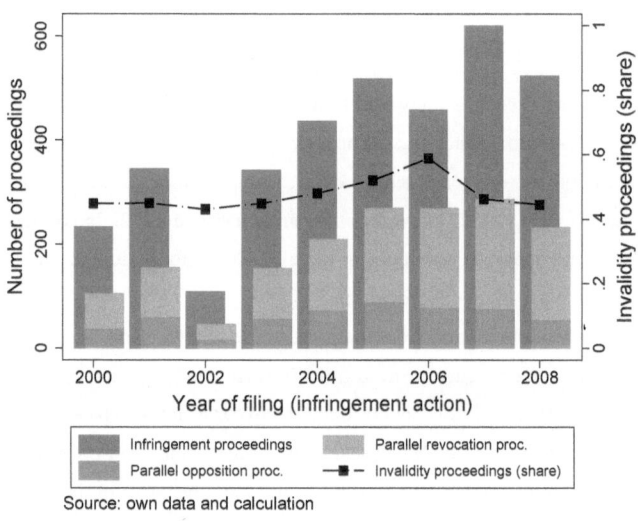

Source: own data and calculation

Figure 3.3 shows the number of infringement and revocation proceedings as well as oppositions by filing year at a patent case level. There has been an increase in caseload over time.[84] The figure also shows that the majority of parallel validity challenges are revocation actions filed before the BPatG. Only 30.8% of all validity challenges are oppositions. The share of infringement cases with a parallel invalidity proceeding is around 48.7% over the entire 2000 to 2008 period.

---

[82]As parallel revocation proceedings may be filed either before or after this time frame, we identified all revocation actions filed against patents involved in an infringement proceeding and added these to our data. For more details and a breakdown of cases by court, see Cremers *et al.* (2013).

[83]For oppositions filed against *DE* patents at the DPMA, we obtained only data for our sample of infringement patents.

[84]The dip in 2002 is due to an internal decision at the regional court in Düsseldorf to remove and destroy files and only store decisions in the court archive. Fortunately, this decision affected only our data for 2002.

3.5 Results

3.5.1 Divergent Decisions

First, we assess the frequency of divergent decisions where the patent was 'invalid but infringed.' One of the factors mentioned in Section 3.3.1 that contributes to the occurrence of divergent decisions is the temporal gap between infringement and validity challenges. Figure 3.4 shows the time lags between the filing of infringement and revocation actions. The distribution shows that in most cases the revocation action followed its corresponding infringement action. As revocation proceedings take longer on average (see Figure 3.7 in the Appendix), the infringement decision is usually handed down first, despite the possibility of staying the infringement proceeding. Figure 3.8 in the Appendix shows the time lags between infringement actions and oppositions. Here, oppositions are largely initiated before the infringement action, suggesting that oppositions are used preemptively rather than reactively. Still, the (first instance) decision on validity is rendered on average 6.7 months *after* the (first instance) decision on infringement. This shows there is substantial scope for an injunction gap; i.e., if a patent is found to be infringed, the patent holder has on average 6.7 months to enforce the patent until the patent is potentially invalidated.

Figure 3.4: Timing of infringement and revocation actions in parallel proceedings

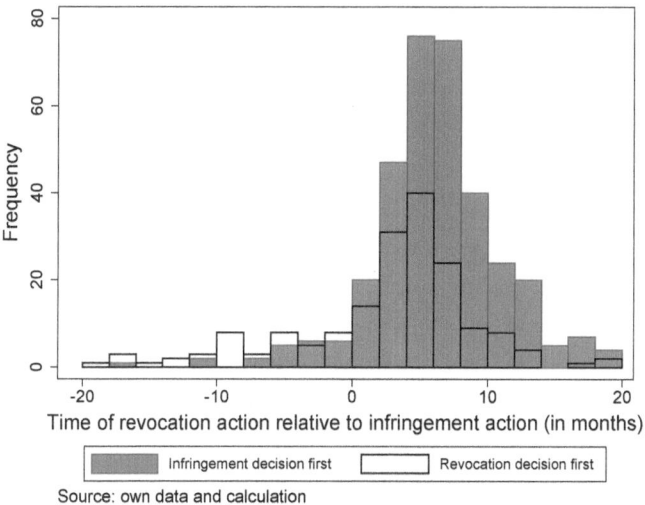

Source: own data and calculation

In Table 3.1 we cross-tabulate the (first instance) infringement and invalidity outcomes for all 1,240 parallel cases where the decision on infringement was handed down first.[85] The gray-shaded cells in Table 3.1 show that there is a sizable number of cases where the patent was first found to be infringed and later invalidated by the BPatG or the DPMA/EPO. If we also consider cases where the patent was partly invalidated or infringed, there is a total of 144 cases. For comparison, patents found to be (partly) infringed were upheld in the invalidity proceeding in 79 cases. This means that slightly more than 12.3% of cases (including cases that settled) produced divergent decisions – the patent was first found to be infringed but later invalidated.[86] If we focus on cases with a decision in both venues, the share increases to 41.8%. We also observe 183 cases where the patent was found (partly) infringed and the parallel invalidity proceeding subsequently ended with a withdrawal of the action.[87]

Table 3.1: Outcomes of infringement and invalidity proceedings where infringement was decided first

| Outcome LG | Outcome parallel invalidity proceeding | | | | |
	valid	partly invalid	invalid	withdrawn	Total
infringed	58	50	55	132	296
	43.6%	24.0%	25.1%	19.4%	23.8%
partly infringed	21	21	18	51	111
	15.8%	10.1%	8.2%	7.5%	9.0%
not infringed	23	45	58	87	213
	17.3%	21.6%	26.5%	12.8%	17.2%
settlement	31	92	88	410	621
	23.3%	44.2%	40.2%	60.3%	50.1%
Total	133	208	219	680	1,240
	100.0%	100.0%	100.0%	100.0%	100.0%

Notes: Dark gray-shaded area shows clear divergent decisions. Light gray-shaded area shows presumed divergent decisions. The sample consists of all infringement proceedings with a parallel invalidity proceeding and where the first instance infringement outcome came first. In case of multiple invalidity decisions, the fastest decision is chosen. The unit of observation is the patent in the infringement proceedings.

Figure 3.5 shows the length of the injunction gap for the 144 cases with divergent decisions. The figure distinguishes between invalidity decisions through the opposition divisions of the DPMA/EPO and the BPatG. The median injunction gap for cases in which the infringed patent was eventually invalidated by the BPatG is approximately 14 months. Hence, parties that have won the infringement case have a little over a year to enforce a patent that should not have been granted in the first place.[88] The length of the injunction gap is significantly longer for cases in which the patent was invalidated through opposition procedures. The median is 34

---

[85]Table 3.7 in the Appendix shows the cases where invalidity was decided first.

[86]Figure 3.9 in the Appendix shows the occurrence of divergent decisions over time.

[87]Figure 3.10 in the Appendix shows that the majority of these cases ended in a settlement shortly after the infringement decision.

[88]Considering that appellate invalidity proceedings take several years, the actual injunction gap until the decision on the patent's invalidity is binding may be considerably longer.

Figure 3.5: Length of injunction gap for divergent decisions

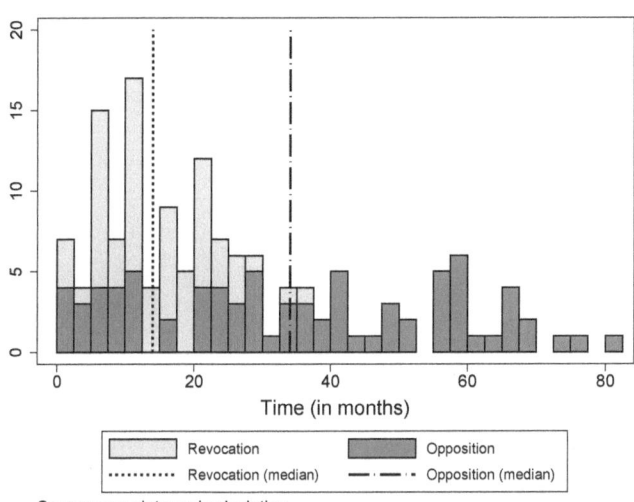

Source: own data and calculation

Notes: The figure shows all divergent decisions regardless of whether parties have (preliminarily) enforced the infringement judgment.

months. As shown in Figure 3.5, the main reason for this is that a considerable number of opposition proceedings take a lot longer to reach a final decision than revocation proceedings at the BPatG.

As explained in Section 3.2, the judgment on infringement is (preliminarily) enforceable despite a pending decision on validity. The only way to prevent an injunction from taking binding effect is to appeal the judgment. In fact, we observe an appeal rate of 54.0% for cases with a parallel invalidity proceeding compared to 26.2% for cases with no parallel invalidity proceeding (see Table 3.8 in the Appendix).[89] Table 3.2 presents the final outcomes of cases with divergent decisions (Figure 3.11 in the Appendix shows a more detailed breakdown). The table shows that the divergent decision was upheld on appeal in 43.1% of the cases. In 13 cases, the court upheld the patent on appeal, which means the defendant in the infringement proceeding is indeed infringing a (partly) valid patent.[90] It bears mentioning that a substantial number of cases are settled during appeal. It is difficult to interpret these numbers, but the fact

---

[89]This suggests that the appeal rate is higher in a bifurcated system than in a non-bifurcated system. An appeal to an infringement decision may be useful – regardless of its prospects of success – as a way to delay an injunction while the decision on validity is still pending.

[90]We only have incomplete information on the timing of the infringement decision in second instance relative to the invalidity decision. It may be that the appeals court did not wait until the decision on validity was available, although infringement proceedings are more likely to be stayed in second instance.

that the share of settled infringement cases (38.2%) is more than three times as large as the share of settled invalidity cases (11.8%) suggests that firms that have been found to infringe a patent are prone to settle (presumably on terms favorable to the patent holder).

Table 3.2: Final outcome to divergent decisions

Final infringement outcome	Final validity outcome			Total
	reversed	settled	binding	
reversed	2	0	6	8
	1.4%	0.0%	4.2%	5.6%
settled	8	11	36	55
	5.6%	7.6%	25.0%	38.2%
binding	13	6	62	81
	9.0%	4.2%	43.1%	56.3%
Total	23	17	104	144
	16.0%	11.8%	72.2%	100.0%

Notes: The sample consists of all infringement proceedings with a divergent decision. The unit of observation is the patent in each infringement proceeding. The observable outcome of oppositions is by definition binding. Settlements are broadly defined and include withdrawn appeals.
White area: divergent decisions eventually reversed by the respective appeals court.
Light gray-shaded area: divergent decisions where at least one appeal proceeding ended with a settlement.
Dark gray-shaded area: divergent decisions that remained unaltered due to lack of appeal or an affirmative decision by the appeals court.

Table 3.3 shows a comparison of case, litigant, and patent level characteristics between defendants in infringement cases subject to divergent decisions and all other cases with non-divergent outcomes. The litigation value does not differ significantly between divergent and non-divergent decisions, i.e., there is no evidence for disproportionately many low-value cases ending up in an injunction gap. That said, we find that validity challenges in cases with divergent decisions are filed on average two and a half months later than in cases with non-divergent decisions. This can be interpreted in different ways. The infringement court may reject requests for a stay more often if the validity challenge has been filed with considerable delay. Alternatively, longer preparation time for a validity challenge may increase the chance of success. In any case it shows that the temporal separation of infringement and invalidity proceedings contributes to divergent outcomes. Looking at the size of the alleged infringers in the two groups, we observe more small firms in the divergent decision group and more large firms in the non-divergent decision group. We also see a larger share of patents belonging to the instruments technology area among divergent decisions. Most of these cases involve patents on medical device technologies. At the same time, there are a lot fewer cases based on patents from the electrical engineering technology area. Interestingly, we do not observe a significant difference in the representation of top law firms before court.

Table 3.3: Comparison of alleged infringers by decision

| Variables | Decision type | | | |
| | Non-divergent | Divergent | | |
	Mean	Mean	SEM	Significance level
Alleged infringer				
Micro	0.11	0.13	0.029	
Small	0.14	0.25	0.032	**
Medium	0.22	0.25	0.037	
Large	0.52	0.38	0.045	**
Germany	0.85	0.88	0.031	
Europe (excl. Germany)	0.10	0.08	0.026	
World (excl. Europe)	0.05	0.04	0.019	
Top legal representative	0.55	0.63	0.044	
Proceeding				
Litigation value (in th €)	1,137.70	992.24	231.247	
Lag of revocation action (in months)	4.16	6.57	0.919	**
Lag of opposition (in months)	-6.70	-6.75	1.544	
Technology area				
Electrical engineering	0.30	0.15	0.040	***
Instruments	0.12	0.22	0.030	**
Chemistry	0.15	0.20	0.032	
Mechanical engineering	0.28	0.28	0.040	
Other	0.15	0.15	0.032	
Observations	1,096	144		

* $p<0.05$, ** $p<0.01$, *** $p<0.001$

Notes: The sample consists of all infringement proceedings with parallel revocation proceedings or oppositions regardless of the timing of the decisions. The unit of observation is the patent in each infringement proceeding. SEM: standard error of mean difference.

Figure 3.3 in Section 3.4.2 showed that about 40% of infringement cases (counted at the patent level) are associated with a revocation action or opposition. This figure is low compared to litigation systems where infringement and invalidity are decided simultaneously in the same proceeding. In non-bifurcated litigation systems, a counterclaim for invalidity is a standard defense to alleged infringement. Therefore, the low figure in our German data is a sign of self-selection among litigants.

The model in Section 3.3 suggested that in a bifurcated system more resource-constrained parties are less likely to contest validity. We test this hypothesis by estimating the propensity that the alleged infringer files a revocation action at the BPatG. That is, we predict the probability that the validity of a patent involved in an infringement case is also challenged at the BPatG. We include a number of patent, case, and litigant characteristics among the regressors (for summary statistics see Table 3.8 in the Appendix). The regressions also include year, patent technology class, and court dummies. Table 3.4 shows the results.

Table 3.4: Probit model results: incidence of revocation action

	(A1) Action filed		(A2) Action filed		(A3) Action filed		(A4) Action filed	
Alleged infringer								
Small (d)	0.081	(0.044)	0.078	(0.044)	0.060	(0.044)	0.049	(0.045)
Medium (d)	0.107**	(0.040)	0.104*	(0.040)	0.092*	(0.041)	0.090*	(0.042)
Large (d)	0.158***	(0.036)	0.175***	(0.037)	0.135***	(0.039)	0.120**	(0.040)
Europe (excl. Germany) (d)					−0.095***	(0.026)	−0.095***	(0.026)
World (excl. Europe) (d)					−0.240***	(0.035)	−0.228***	(0.037)
Top legal represent. (JUVE) (d)							0.136***	(0.021)
Patent holder								
Small (d)			0.006	(0.048)	−0.009	(0.048)	−0.028	(0.048)
Medium (d)			−0.029	(0.045)	−0.033	(0.046)	−0.047	(0.046)
Large (d)			−0.006	(0.042)	−0.011	(0.043)	−0.025	(0.044)
Europe (excl. Germany) (d)					0.040	(0.027)	0.018	(0.027)
World (excl. Europe) (d)					−0.068*	(0.030)	−0.057	(0.032)
Top legal represent. (JUVE) (d)							0.024	(0.028)
Patent characteristics								
Forward citations (in first 5 years)	−0.001	(0.002)	0.002	(0.002)	−0.002	(0.002)	−0.003	(0.003)
EP bundle patent (d)					−0.034	(0.028)	−0.062*	(0.030)
Acc. examination requested (d)							0.094**	(0.033)
Grant lag (diff. from mean in days)							−0.000*	(0.000)
Age of patent (in years)			−0.003	(0.002)	0.007	(0.012)	0.018	(0.012)
Age of patent (in years, squared)					−0.000	(0.001)	−0.001	(0.001)
Invalidity history								
Patent solidified (opp. proc.) (d)	0.054	(0.028)	0.069*	(0.028)	0.034	(0.029)	0.014	(0.029)
Patent challenged (rev. proc.) (d)	−0.191***	(0.031)	−0.163***	(0.039)	−0.191***	(0.039)	−0.170***	(0.042)
Patent solidifed (rev. proc.) (d)			−0.043	(0.077)	0.003	(0.081)	−0.033	(0.080)
Proceeding								
Parallel opposition proceeding (d)			−0.214***	(0.036)	−0.243***	(0.033)	−0.265***	(0.031)
Litigation value (in th €)			−0.000	(0.000)				
Litigation value (in th €, log)					0.039***	(0.010)	0.042***	(0.010)
Multi-jurisdictional litigation (d)							0.123*	(0.058)
Controls								
Year effects	Yes***		No		No		Yes***	
Technology effects	Yes***		Yes***		Yes**		Yes**	
Court effects	No		Yes		Yes		Yes**	
Patent characteristics	Yes**		No		Yes***		Yes***	
Pseudo R^2	0.039		0.037		0.065		0.096	
Observations	2,397		2,388		2,388		2,388	

Marginal effects; Standard errors in parentheses;
(d) for discrete change of dummy variable from 0 to 1; * $p < 0.05$, ** $p < 0.01$, *** $p < 0.001$
Notes: The sample consists of all infringement proceedings with a duration of at least 120 days. The unit of observation is at the patent-case level, i.e., each patent in each infringement proceeding is treated as a separate case. Baseline patent holder size: micro. Baseline alleged infringer size: micro.

Our focus is on the size of the alleged infringers, who have to decide whether to challenge the validity of the allegedly infringed patents before the BPatG. We distinguish between four size categories: micro, small, medium and large.[91] The results for our preferred specification in Column (A4) show that medium-sized and large firms are about 12% more likely to file a revocation action at the BPatG than micro-sized alleged infringers. This suggests that after accounting for time varying and invariant patent and case characteristics, smaller defendants in infringement proceedings are less likely to challenge the validity of the patent at issue. In contrast, there is no evidence that the size of the plaintiff in the infringement proceedings plays any role in the decision to challenge the patent's validity. This supports the view that the decision *not* to file a parallel action at the BPatG is at least partly determined by resource constraints on the alleged infringer's side.

The results also show that – in line with the empirical prediction in Section 3.3.1 – firms not registered in Germany are less likely to challenge a patent's validity. Firms in a European country other than Germany are about 9% less likely to file a parallel action, and firms outside of Europe are approximately 23% less likely than German firms to file a revocation action. This may be explained by the larger costs involved for parties outside of Germany in pursuing an additional action at the BPatG. For example, since the court action is conducted in German, translation costs accrue, and in most cases representation has to be assigned to a German law firm. In line with this reasoning, we find that cases in which the alleged infringer is represented by a top law firm to be 14% more likely to have a parallel invalidity proceeding.

To further explore this screening effect of bifurcation, we use propensity score matching to create a weighted sample more similar on the covariates. We estimate the propensity for a parallel revocation action to be pursued at the BPatG by a small or large defendant in an infringement case. In contrast to Table 3.4, for the propensity matching estimation, we collapse the data into two firm size categories: micro/small and medium/large. The results in Table 3.5 echo the probit results of Table 3.4.[92] Large and medium-sized firms have a higher likelihood of pursuing a revocation action than micro and small firms, ceteris paribus. This result holds regardless of the way we match treated and control units (propensity score or nearest neighbor) and the number of matched controls (1 or 5). The lower part of Table 3.5 also shows results when excluding non-European alleged infringers. The average treatment effects are slightly larger in magnitude but overall very similar to the results obtained for the full sample.

These results provide strong evidence that bifurcation deters validity challenges, as sug-

---

[91]The size categories are defined according to the EU definition available at http://ec.europa.eu/enterprise/policies/sme/facts-figures-analysis/sme-definition/index_en.htm [accessed: 22 July 2015], which relies on information on a firm's number of employees, turnover, and total assets.

[92]Figure 3.12 in the Appendix shows that treated and control units have common support.

Table 3.5: Average treatment effects

Matches per observation	Propensity Score Matching		Nearest Neighbor Matching	
	nn=1	nn=5	nn=1	nn=5
Full sample (N=2,376)				
ATE action filed	0.09	0.08	0.11	0.12
Std. err.	0.02	0.02	0.03	0.03
P-value	0.000	0.000	0.000	0.000
Matches minimum	1	5	1	5
Matches maximum	4	12	4	7
European alleged infringers (N=2,262)				
ATE action filed	0.11	0.08	0.12	0.13
Std. err.	0.02	0.01	0.03	0.03
P-value	0.000	0.000	0.005	0.000
Matches minimum	1	5	1	5
Matches maximum	4	7	4	15

Notes: The sample consists of all infringement proceedings with a duration of at least 120 days. The unit of observation is at the patent-case level, i.e., each patent in each infringement proceeding is treated as a separate case. Cases with a patent belonging to a technological class with fewer than five patents in the entire sample are excluded. ATE: average treatment effect. Treatment model: logit. Distance metric: Mahalanobis.

gested by our model. More resource-constrained firms are less likely to file a revocation action in response to an infringement claim. This also implies that the 12% of divergent decisions shown in Section 3.5.1 are downward biased. Fewer patents are invalidated than in the absence of the additional costs caused by the separation of proceedings. A fact that contributes to the strong presumption of validity in a self-enforcing way.

3.5.2 Effect on Oppositions

It is difficult to empirically gauge the effect a divergent decision has on alleged infringers. If the infringement decision is indeed enforced while the decision on validity is pending, it is reasonable to expect some direct negative effect. In addition, firms may also adjust their expectations about facing an injunction despite the expected invalidity of a patent. Such uncertainty about the likelihood of infringing a patent right may affect a firm's behavior beyond the immediate direct effect of the injunction. We therefore analyze whether firms try to avoid repeated exposure to divergent decisions by adjusting their opposition behavior.

In fact, there is reason to believe that a bifurcated litigation system increases a firm's incentives to reduce uncertainty by attempting to eliminate patents early on through oppositions. Figure 3.13 in the Appendix shows that German firms are responsible for more than half of all opposition proceedings before the EPO between 1997 and 2013. German firms oppose disproportionately more patents at the EPO than firms from other countries.

Oppositions are a relatively cheap and effective means of clearing potentially harmful patents early on. Once they become subject to a divergent decision, firms may file more oppositions against patents to preempt the risk of future infringement allegations. We test for an increase in opposition filings by conducting an event study analysis: we regress the number of oppositions by a firm that has faced an injunction gap on a dummy variable that equals one once the decision on validity is handed down and it becomes apparent that the regional court had held an invalid patent infringed (diff-specification). We use a ±6 month window to assess changes in opposition behavior. To account for any general tendency for firms to change their opposition behavior following an infringement dispute, we match the set of firms subject to divergent decisions to the set of firms that filed a validity challenge but were not subject to divergent decisions. The interaction term of the dummy variable indicating whether a firm was subject to a divergent decision and the dummy variable indicating the timing of the decision provides a differences-in-differences estimate of any effect on firms' opposition activity (diff-in-diff specification).

Table 3.6 shows the main results. We find a positive coefficient for the interaction term as well as for the post-invalidity dummy. This suggests that firms accused of patent infringement generally increase their opposition behavior following the end of the parallel validity proceeding. This increase in oppositions is, however, stronger for firms that were found to have infringed an invalid patent. The results in Table 3.6 focus on a ±6 month window; Table 3.9 in the Appendix also shows differences in means before and after a case is decided for ±2 and ±12 month windows. The figures for the alternative event windows are consistent with the data shown for the ±6 month window.

These findings are in line with Adam and Spence (2001), who argue that the disproportionate share of oppositions at the EPO by German firms is due to the need to preempt infringement actions in the German bifurcated patent litigation system.[93] Our results, therefore, suggest that the relatively large number of oppositions by German firms is at least in part a manifestation of the uncertainty created by bifurcation.

---

[93]There are two alternative, although not mutually exclusive explanations for the observed increase in opposition activity (Harhoff, 2005). First, oppositions serve to not only preempt specific infringement allegations, but also to develop a reputation for toughness. Second, the alleged infringer may have obtained information (e.g., prior art) during the invalidity proceeding that can be used as evidence against other patents as well. Note that because in our data the alleged infringer does not file a disproportionate number of oppositions against the patent holder of the infringement proceeding, retaliation is an unlikely reason for the increase in opposition activity.

Table 3.6: Differences-in-differences results: oppositions pre/post-invalidity decision

	diff-specification		diff-in-diff specification	
	(1) Oppositions	(2) Oppositions	(3) Oppositions	(4) Oppositions
±6 months				
Post-invalidity decision	0.377***(0.099)	1.504***(0.434)	0.520***(0.155)	0.522***(0.155)
Post-inval. *x* infringed			1.098** (0.387)	1.099** (0.387)
No. of filed patents				−0.000 (0.002)
Opponent fixed effects	Yes	Yes	Yes	Yes
Time effects	No	No	Yes	Yes
Opponents w/ divergent decision	20	20	20	20
Opponents w/ non-divergent decision	95	0	95	95
Observations	690	120	690	690

Standard errors in parentheses; * $p < 0.05$, ** $p < 0.01$, *** $p < 0.001$

Notes: Fixed effect negative binomial regression. Dependent variable number of oppositions filed by alleged infringer. Date of invalidity decision defined as publication of decision by the BPatG (for revocation proceedings) and the DPMA/EPO (for oppositions). Alleged infringers with no oppositions after invalidity decision excluded.

3.6   Conclusion

Our theoretical and empirical results suggest that the bifurcated litigation system strongly favors the patent holder in litigation. We show that this occurs for two reasons: first, the bifurcated system creates a substantial number of cases where an invalid patent is held infringed, and second, fewer patents are challenged than we would expect based on the consideration of litigation systems where infringement and validity are dealt with jointly. This also means that the number of divergent decisions is presumably biased downwards due to self-selection. Our results indicate that the possibility of facing an injunction based on an invalid patent creates legal uncertainty. We find evidence for such legal uncertainty by looking at changes in firms' opposition behavior directly following a divergent decision. We find that alleged infringers who are subject to a divergent decision file more oppositions immediately afterwards.

Our results provide both theoretical and empirical support for the criticism directed at bifurcation (Hilty and Lamping, 2011; Münster-Horstkotte, 2012). Given the probabilistic nature of patents, the strong presumption of validity that sits at the core of Germany's bifurcated patent litigation system favors the patent holder and creates uncertainty for potential infringers. That said, the problems revealed by our analysis should be compared to possible benefits of bifurcation – in particular, the impact of technically trained judges at the BPatG on validity decisions and the lower costs of litigation due to a reduced need for technical expert opinions. Indeed, the costs of litigation in Germany are remarkably low compared to other jurisdictions, such as the U.K.[94] However, technically trained judges could also be incorporated in a court system in which validity and infringement are decided jointly, as is the case in Switzerland (Ann, 2011). Similarly, the current design of the Unified Patent Court foresees the appointment of technically qualified judges if infringement and validity are heard by the same local or regional division.

It is also possible that reforms that increased caseloads at the BPatG and its appeal court, BGH, aggravated the problems during our period of analysis. Regardless, our results suggest that the current system is in danger of overly favoring patent holders. One way of addressing the problems highlighted by our analysis could be an acceleration of proceedings at the BPatG, thereby either avoiding the injunction gap altogether or cutting its length. This could retain the benefits of bifurcation while avoiding the uncertainty created by divergent decisions. Alternatively, bifurcation could be abandoned altogether. While such a drastic step may seem appealing to some observers, its impact would be uncertain. In any case, we do not present a counterfactual analysis that would allow us to evaluate such a drastic step. Our analysis still suggests that bifurcation has considerable drawbacks, at least the way it is currently implemented in the German patent system.

---

[94]Low litigation costs are also a characteristic of the French and the Dutch patent litigation systems, which are not bifurcated (Cremers et al., 2013).

3.7   Appendix to Chapter 3

3.7.1   Figures

Figure 3.6: Court structure in Germany's patent system (Cremers *et al.*, 2013, amended)

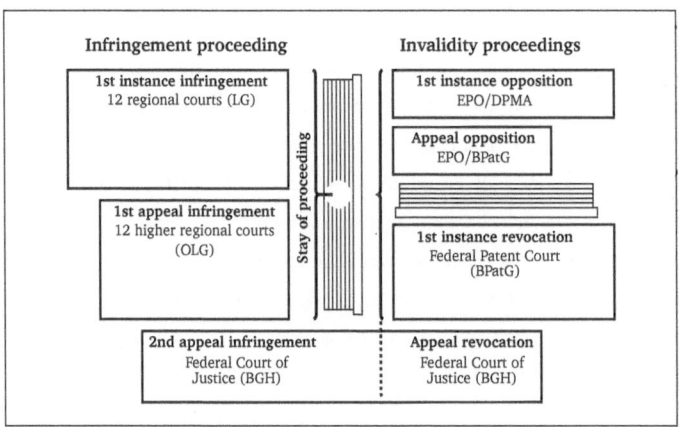

Figure 3.7: Length of (first instance) revocation proceedings by year

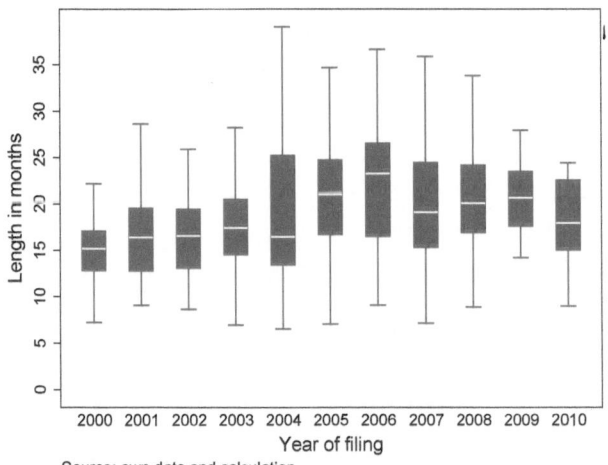

Figure 3.8: Timing of infringement and oppositions in parallel proceedings

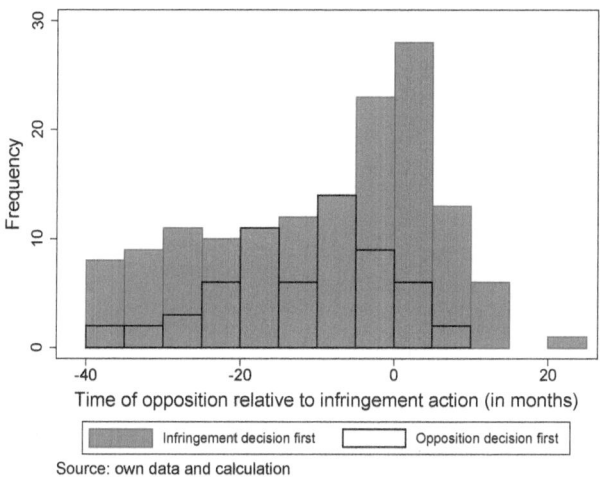

Source: own data and calculation

Figure 3.9: Number and share of divergent decisions over time

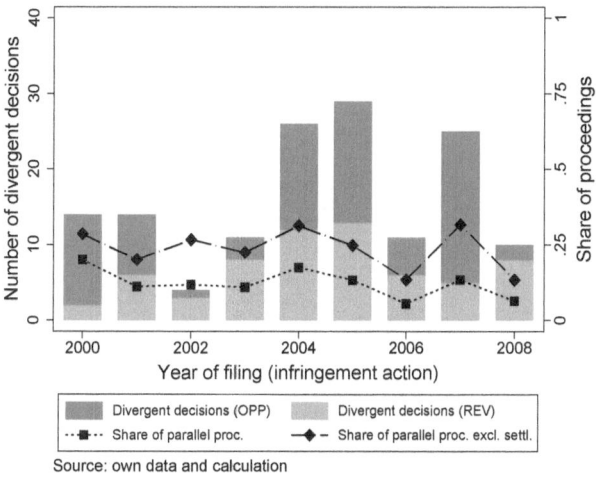

Source: own data and calculation

Notes: Only parallel proceedings with outcome in infringement proceeding first. Share of parallel proceedings excluding settlements includes settlements in the infringement but not invalidity proceeding.

Figure 3.10: Time between infringement decision (first instance) and settlement in parallel invalidity proceeding

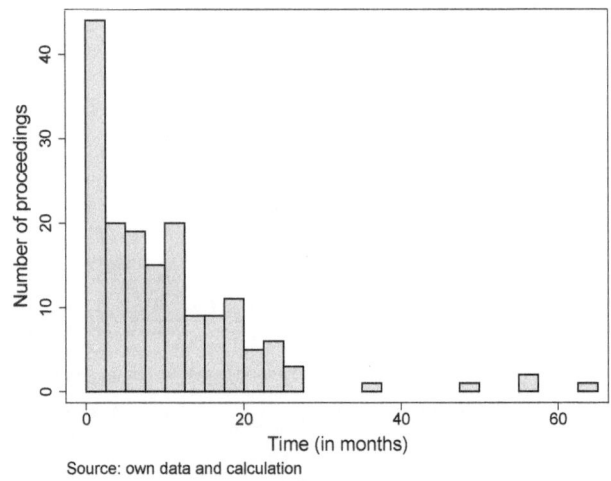

Source: own data and calculation

Notes: Only parallel proceedings with outcome 'infringed' or outcome 'partly infringed'.

Figure 3.11: Appeals and settlements of divergent decision cases

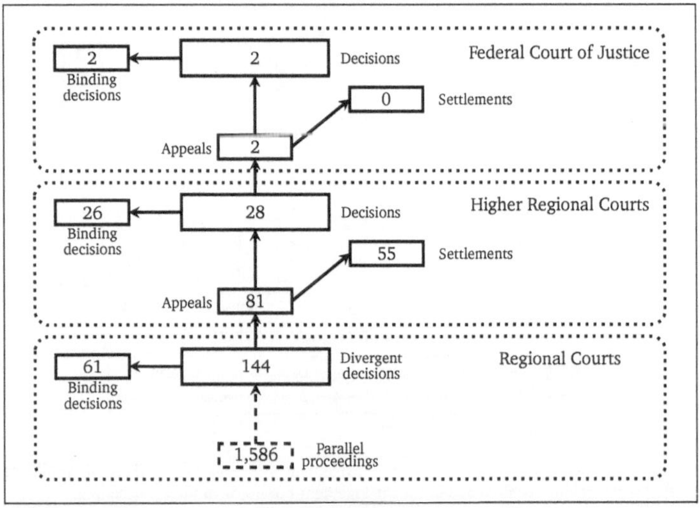

```
┌────────────────────────────────────────────────────────────────────────┐
│  ┌ ─ ─ ─ ─ ─ ─ ─ ─ ─ ─ ─ ─ ─ ─ ─ ─ ─ ─ ─ ─ ─ ─ ─ ─ ─ ─ ─ ─ ─ ─ ─ ─┐     │
│      ┌───┐      ┌──────────────┐                                          │
│      │ 2 │◄──── │      2       │  Decisions    Federal Court of Justice   │
│      └───┘      └──────────────┘        ┌───┐                             │
│      Binding                            │ 0 │  Settlements                │
│      decisions                          └───┘                             │
│                          Appeals ┌───┐                                    │
│                                  │ 2 │                                    │
│  └ ─ ─ ─ ─ ─ ─ ─ ─ ─ ─ ─ ─ ─ ─ ─└───┘─ ─ ─ ─ ─ ─ ─ ─ ─ ─ ─ ─ ─ ─ ─┘     │
│  ┌ ─ ─ ─ ─ ─ ─ ─ ─ ─ ─ ─ ─ ─ ─ ─ ─ ─ ─ ─ ─ ─ ─ ─ ─ ─ ─ ─ ─ ─ ─ ─ ─┐     │
│      ┌────┐     ┌──────────────┐                                         │
│      │ 26 │◄────│      28      │  Decisions     Higher Regional Courts    │
│      └────┘     └──────────────┘        ┌────┐                           │
│      Binding                            │ 55 │  Settlements              │
│      decisions                          └────┘                           │
│                          Appeals ┌────┐                                   │
│                                  │ 81 │                                   │
│  └ ─ ─ ─ ─ ─ ─ ─ ─ ─ ─ ─ ─ ─ ─ ─└────┘─ ─ ─ ─ ─ ─ ─ ─ ─ ─ ─ ─ ─ ─ ─┘     │
│  ┌ ─ ─ ─ ─ ─ ─ ─ ─ ─ ─ ─ ─ ─ ─ ─ ─ ─ ─ ─ ─ ─ ─ ─ ─ ─ ─ ─ ─ ─ ─ ─ ─┐     │
│      ┌────┐     ┌──────────────┐  Divergent                             │
│      │ 61 │◄────│     144      │  decisions      Regional Courts          │
│      └────┘     └──────────────┘                                         │
│      Binding                                                             │
│      decisions      ┌ ─ ─ ─ ─ ┐                                          │
│                     │  1,586  │  Parallel                                │
│                     └ ─ ─ ─ ─ ┘  proceedings                             │
│  └ ─ ─ ─ ─ ─ ─ ─ ─ ─ ─ ─ ─ ─ ─ ─ ─ ─ ─ ─ ─ ─ ─ ─ ─ ─ ─ ─ ─ ─ ─ ─ ─┘     │
└────────────────────────────────────────────────────────────────────────┘
```

Figure 3.12: Distribution of estimated propensity scores

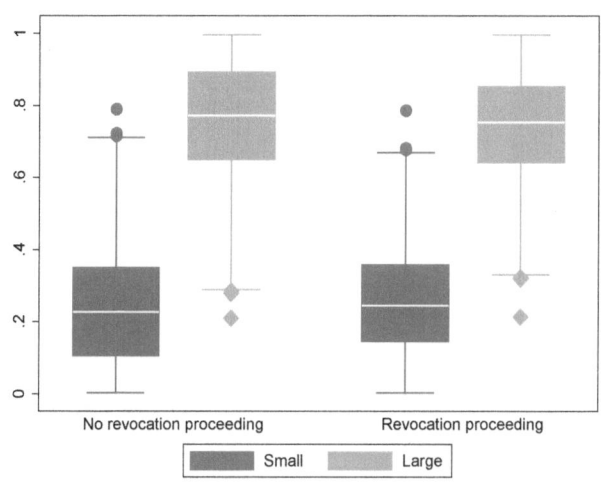

Figure 3.13: Oppositions filed against *EP* patents by year

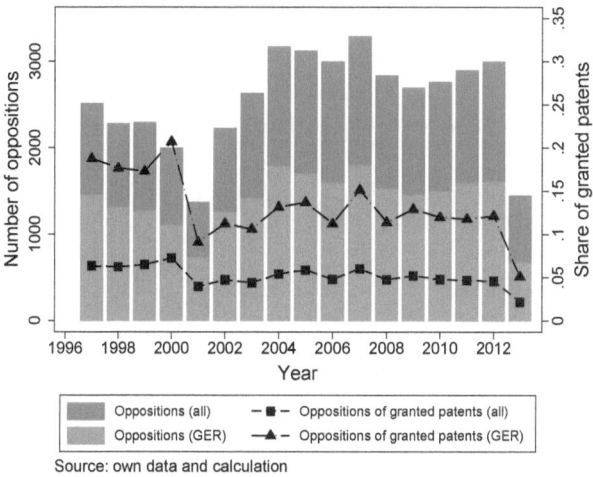

Notes: Figures on granted patents by country and year according to the annual reports of the EPO. GER represents the subsample of all granted patents (filed oppositions) with the first applicant (opponent) originating from Germany. Oppositions are counted on patent level. Data for oppositions filed in 2013 truncated.

3.7.2   Tables

Table 3.7: Outcomes of infringement and invalidity proceedings where validity was decided first

	Outcome parallel invalidity proceeding				
Outcome LG	valid	partly invalid	invalid	withdrawn	Total
infringed	24	15	5	47	91
	33.3%	16.9%	3.9%	29.9%	20.4%
partly infringed	9	8	3	2	22
	12.5%	9.0%	2.3%	1.3%	4.9%
not infringed	9	18	32	8	67
	12.5%	20.2%	24.8%	5.1%	15.0%
settlement	30	48	89	100	267
	41.7%	53.9%	69.0%	63.7%	59.7%
Total	72	89	129	157	447
	100.0%	100.0%	100.0%	100.0%	100.0%

Notes: Gray-shaded areas show divergent decisions. The sample consists of all infringement proceedings with a parallel invalidity proceeding and where the first instance infringement outcome is first. In case of multiple invalidity decisions, the fastest decision is chosen. The unit of observation is the patent in the infringement proceedings. Occurrence of divergent decisions explained by multiple patents in same proceeding, appeal to invalidity decision, or missing defense by alleged infringer.

Table 3.8: Summary statistics grouped by parallel revocation proceeding

	No parallel revocation proceeding				Parallel revocation proceeding			
Variables	Mean	Std. dev.	Min	Max	Mean	Std. dev.	Min	Max
Patent characteristics								
EP bundle patent (d)	0.77	0.42	0	1	0.78	0.42	0	1
PCT filing (d)	0.21	0.41	0	1	0.19	0.39	0	1
Forward citations (in first 5 years)	3.30	4.93	0	44	5.32	6.18	0	51
Backward citations (patents)	4.74	2.89	0	27	5.42	3.44	0	32
Backward citations (nonpatent literature)	0.83	1.97	0	21	1.65	2.47	0	18
IPC subclass count	2.24	1.90	1	9	3.32	2.94	1	9
Family size (INPADOC)	11.61	16.39	1	183	20.62	24.75	1	69
Year of patent application/priority	1992.50	4.88	1980	2004	1992.28	4.37	1979	2005
Grant lag (diff. from mean in days)	69.56	726.10	-1,193	4,641	21.06	601.81	-1,303	4,004
Acc. examination requested (d)	0.12	0.33	0	1	0.13	0.34	0	1
Age of patent (in years)	12.18	4.78	1	25	12.54	4.49	1	23
Patent technology area								
Chemistry (d)	0.12	0.32	0	1	0.13	0.33	0	1
Electrical engineering (d)	0.32	0.47	0	1	0.38	0.49	0	1
Instruments (d)	0.10	0.30	0	1	0.12	0.33	0	1
Mechanical engineering (d)	0.29	0.45	0	1	0.23	0.42	0	1
Other (d)	0.17	0.38	0	1	0.14	0.35	0	1

Continued on next page

Table 3.8 – continued from previous page

Variables	No parallel revocation proceeding				Parallel revocation proceeding			
	Mean	Std. dev.	Min	Max	Mean	Std. dev.	Min	Max
Patent invalidition history								
Patent solidifed (opp. proc.) (d)	0.14	0.35	0	1	0.16	0.37	0	1
Patent challenged (rev. proc.) (d)	0.14	0.35	0	1	0.26	0.44	0	1
Patent solidifed (rev. proc.) (d)	0.03	0.18	0	1	0.01	0.11	0	1
Infringement proceeding								
Year of infringement action (d)	2004.77	2.53	2000	2008	2004.93	2.35	2000	2008
Litigation value (in thousand €)	1,292.62	3,021.39	0	38,348	1,082.32	2,807.63	0	35,000
Length of proceeding (in months)	10.79	11.07	0	128	17.43	16.65	0	128
Parallel opposition proceeding (d)	0.04	0.20	0	1	0.02	0.15	0	1
Multijurisdictional litigation (d)	0.03	0.16	0	1	0.04	0.20	0	1
LG Düsseldorf (d)	0.68	0.47	0	1	0.58	0.49	0	1
LG Mannheim (d)	0.23	0.42	0	1	0.36	0.48	0	1
LG Munich (d)	0.09	0.29	0	1	0.06	0.24	0	1
Judgment appealed (d)	0.26	0.44	0	1	0.54	0.50	0	1
Patent holder								
Nonpracticing entity (d)	0.19	0.39	0	1	0.29	0.45	0	1
Micro (d)	0.13	0.34	0	1	0.11	0.31	0	1
Small (d)	0.10	0.29	0	1	0.09	0.28	0	1
Medium (d)	0.16	0.36	0	1	0.13	0.34	0	1
Large (d)	0.61	0.49	0	1	0.67	0.47	0	1
Germany (d)	0.69	0.46	0	1	0.63	0.48	0	1
Europe (excl. Germany) (d)	0.19	0.39	0	1	0.31	0.46	0	1
World (excl. Europe) (d)	0.12	0.33	0	1	0.06	0.24	0	1
Top legal representative (JUVE) (d)	0.67	0.47	0	1	0.60	0.49	0	1
Alleged infringer								
Micro (d)	0.16	0.37	0	1	0.17	0.38	0	1
Small (d)	0.17	0.38	0	1	0.14	0.35	0	1
Medium (d)	0.25	0.43	0	1	0.23	0.42	0	1
Large (d)	0.41	0.49	0	1	0.46	0.50	0	1
Germany (d)	0.72	0.45	0	1	0.83	0.38	0	1
Europe (excl. Germany) (d)	0.18	0.38	0	1	0.11	0.31	0	1
World (excl. Europe) (d)	0.10	0.30	0	1	0.07	0.25	0	1
Top legal representative (JUVE) (d)	0.38	0.49	0	1	0.52	0.50	0	1
N	1,971				1,226			

Table 3.9: Comparison of oppositions by alleged infringers at the EPO

	Mean		SEM
	Before	After	
Opponents with non-divergent decisions			
±2 months			
Filed patents	11.22	13.44	3.214
Filed oppositions	0.22	0.25	0.075
±6 months			
Filed patents	35.18	37.65	9.152
Filed oppositions	0.77	0.80	0.246
±12 months			
Filed patents	73.31	72.84	18.011
Filed oppositions	1.55	1.68	0.499
Opponents with divergent decisions			
±2 months			
Filed patents	7.05	9.46	4.840
Filed oppositions	0.10	0.34	0.128
±6 months			
Filed patents	19.07	22.56	11.638
Filed oppositions	0.27	0.85	0.238
±12 months			
Filed patents	39.95	45.07	23.404
Filed oppositions	1.17	1.41	0.462

Notes: The sample consists of all alleged infringers that have filed at least one opposition against an *EP* patent twelve months before or after a decision in the parallel invalidity proceeding. SEM: standard error of mean difference.

Table 3.10: 'Invalid but infringed' example cases

		First instance					Appeal		Patent		
Proc.	Case number	Filing date	Litigation value in €	Outcome	Outcome date	Case number	Outcome	Outcome date	Publication number	Application date	Technology
INF	DU 4 O 219/00	01-Apr-00	511,300	infringed	10-May-01	2 U 95/01	settlement	12-Apr-07	EP0203206	02-May-85	Chemical engineering
REV	2 Ni 42/00 (EU) et al.	02-Nov-00		invalid	27-Feb-02	X ZR 156/02	aff: invalid	30-Jan-07			
INF	DU 4 O 248/01	19-Jul-01	76,693	infringed	28-May-02	2 U 84/02	n/a	n/a	EP0912130	01-Apr-97	Furniture and games
REV	3 Ni 51/01 (EU)	05-Sep-01		partly invalid	21-Jan-03						
INF	DU 4 O 356/01	07-Jul-01	1,022,583	infringed	31-Mar-05				EP0101552	05-Jul-83	Electrical machinery
REV	2 Ni 63/04 (EU) et al.	01-Dec-04		partly invalid	16-Mar-07						
INF	DU 4A O 233/01	04-Jul-01	500,000	infringed	25-Mar-03	2 U 50/03	settlement	n/a	EP0692562	12-Jul-94	Textile and paper
REV	2 Ni 14/03 (EU)	03-Feb-03		invalid	30-Sep-04	n/a	withdrawn	14-Mar-05			
INF	DU 4A O 234/01	05-Jul-01	511,292	infringed	10-Jan-02	2 U 27/02	settlement	05-Apr-07	EP0646362	20-Sep-94	Medical technology
REV	4 Ni 28/01 (EU)	07-Nov-01		invalid	12-Nov-02	X ZR 16/03	amd: partly invalid	25-Apr-06			
INF	DU 4A O 33/01	26-Dec-00	1,022,600	partly infringed	05-Feb-02	2 U 36/02	settlement	12-Jul-02	EP0548475	26-Sep-92	Other consumer goods
REV	2 Ni 47/01 (EU)	07-Dec-01		invalid	08-May-03	X ZR 115/03	amd: partly invalid	01-Apr-08			
INF	DU 4A O 453/01	17-Jan-02	1,500,000	infringed	14-Jan-03	2 U 25/03	settlement	2-Apr-04	DE3639669	20-Nov-86	Electrical machinery
REV	2 Ni 26/02	16-Aug-02		partly invalid	20-Nov-03	n/a	withdrawn	08-Sep-04			
INF	DU 4A O 185/03	10-May-03	500,000	partly infringed	08-Jul-04	n/a	withdrawn	04-Oct-06	DE3801617	21-Jan-88	Civil engineering
REV	1 Ni 9/04	10-Mar-04		invalid	31-May-05						
INF	DU 4A O 282/03	08-Jul-03	500,000	infringed	20-Jul-04	1-2 U 81/04	settlement	23-Jan-06	EP0947279	25-Feb-99	Machine tools
REV	2 Ni 3/04 (EU)	08-Jan-04		invalid	19-May-05	X ZR 107/05	aff: invalid	30-Jun-09			
INF	DU 4A O 371/03	03-Sep-03	100,000	partly infringed	28-Sep-04	1-2 U 100/04	settlement	n/a	EP0654427	16-Nov-94	Handling
REV	1 Ni 2/04 (EU)	22-Dec-03		invalid	14-Mar-05						
INF	DU 4B O 346/03	01-Sep-03	350,000	infringed	14-Dec-04				EP0291194	26-Apr-88	Measurement
REV	3 Ni 11/01 (EU) et al.	22-Sep-04		invalid	07-Jun-05	X ZR 154/05	amd: partly invalid	05-Nov-08			
INF	DU 4B O 458/03	28-Nov-03	500,000	infringed	14-Oct-04	1-2 U 105/04	settlement	n/a	EP0781234	12-Sep-95	Handling
REV	1 Ni 14/04 (EU)	07-May-04		invalid	07-Mar-06	n/a	withdrawn	22-Apr-10			

Continued on next page

Table 3.10 – continued from previous page

Proc.	First instance					Appeal			Patent		
	Case number	Filing date	Litigation value in €	Outcome	Outcome date	Case number	Outcome	Outcome date	Publication number	Application date	Technology
INF	DU 4A O 152/04	25-Mar-04	500,000	infringed	10-May-05	I-2 U 99/05	settlement	24-Feb-10	EP0706338	16-Jun-94	Furniture and games
REV	4 Ni 64/04 (EU)	14-Dec-04		partly invalid	28-Mar-06						
INF	DU 4A O 453/04	18-Nov-04	100,000	infringed	26-Jul-05	I-2 U 101/05	settlement	27-Oct-11	EP0753420	09-Jul-96	Transport
REV	3 Ni 48/07 (EU) et al.	21-Feb-05		partly invalid	16-May-08	X ZR 75/08	aff: partly invalid	12-Jul-11			
INF	DU 4B O 18/04	04-Nov-03	500,000	infringed	14-Dec-04				EP0291194	26-Apr-88	Measurement
REV	3 Ni 39/04 (EU) et al.	22-Sep-04		invalid	07-Jun-05	X ZR 154/05	amd: partly invalid	05-Nov-08			
INF	DU 4B O 435/04	06-Nov-04	2,000,000	infringed	21-Apr-05	n/a	settlement	n/a	EP0755348	07-Feb-95	Handling
REV	1 Ni 4/04 (EU)	22-Dec-03		invalid	22-Feb-06	X ZR 79/06	withdrawn	26-Aug-10			
INF	DU 4A O 122/05	10-Mar-05	500,000	partly infringed	09-Feb-06	I-2 U 27/06	not infringed	25-Mar-10	EP0361155	06-Sep-89	Civil engineering
REV	2 Ni 38/05 (EU)	02-Aug-05		invalid	29-Oct-07	XA ZR 6/08	amd: partly invalid	02-Apr-09			
INF	DU 4A O 253/05	17-May-05	500,000	infringed	20-Jul-06	I-2 U 90/06	settlement	29-Aug-11	EP0591132	23-Jul-90	Medical technology
REV	4 Ni 40/06 (EU)	22-May-06		invalid	14-Oct-08	n/a	withdrawn	29-Aug-11			
INF	DU 4A O 394/05	31-Aug-05	500,000	infringed	29-Aug-06	I-2 U 115/06	settlement	18-Jul-08	EP0280340	18-Jan-88	Textile and paper
REV	2 Ni 2/06 (EU)	11-Jan-06		partly invalid	21-May-08						
INF	DU 4A O 452/05	11-Sep-05	500,000	infringed	21-Feb-06	I-2 U 25/06	not infringed	6-Sep-07	EP0676763	04-Jul-95	Audio-visual technology
REV	1 Ni 1/06 (EU)	05-Jan-06		invalid	14-Feb-08	XA ZR 85/08	amd: partly invalid	29-Jul-10			
INF	DU 4A O 484/05	04-Nov-05	250,000	infringed	09-May-06	I-2 U 60/06	infringed	27-Sep-06	DE19945719	23-Sep-99	Other special machines
REV	4 Ni 7/06	25-Jan-06		invalid	08-Jan-08	X ZR 49/08	aff: invalid	17-Nov-09			
INF	DU 4A O 552/05	23-Nov-05	500,000	infringed	13-Feb-07	I-2 U 14/07	settlement	4-Nov-10	EP0851376	30-Dec-96	Computer technology
REV	2 Ni 25/06 (EU)	22-Jun-06		invalid	10-Jun-08	n/a	withdrawn	10-Nov-10			
INF	DU 4A O 62/05	15-Feb-05	1,000,000	infringed	09-Mar-06	I-2 U 28/06	settlement	31-Aug-12	EP0821784	18-Apr-96	Measurement
REV	4 Ni 45/05 (EU)	03-Sep-05		partly invalid	02-May-07	XA ZR 84/07	amd: partly invalid	25-Nov-10			
INF	- DU 4B O 128/05	12-Mar-05	500,000	infringed	14-Mar-06	I-2 U 39/06	settlement	27-Aug-10	EP0337612	17-Mar-89	Medical technology
REV	4 Ni 52/05 (EU)	10-Oct-05		invalid	26-Jun-07	XA ZR 126/07	sent back to BPatG	13-Jul-10			

Continued on next page

Table 3.10 – continued from previous page

Proc.	Case number	First instance					Appeal		Patent		
		Filing date	Litigation value in €	Outcome	Outcome date	Case number	Outcome	Outcome date	Publication number	Application date	Technology
INF	DU 4B O 76/05	03-Feb-05	1,000,000	infringed	16-May-06	1-2 U 57/06	settlement	4-Dec-07	EP0350528	15-Jul-88	Thermal processing
REV	1 Ni 8/06 (EU)	27-Apr-06		invalid	20-Mar-07	XA ZR 66/07	amd: partly invalid	14-Jan-10			
INF	DU 4A O 263/06	22-Jul-06	200,000	partly infringed	14-Aug-07	1-2 U 83/07	settlement	19-Aug-09	EP1098706	08-Jul-99	Chemical engineering
REV	3 Ni 48/06 (EU)	30-Jun-06		partly invalid	29-Apr-08	n/a					
INF	DU 4B O 160/06	09-May-06	500,000	partly infringed	22-Jul-08	1-2 U 74/08	settlement	14-Oct-10	DE4337743	05-Nov-93	Civil engineering
REV	3 Ni 77/06	07-Dec-06		invalid	13-Jan-09	n/a	withdrawn	04-Oct-10			
INF	DU 4B O 279/06	15-Jul-06	1,000,000	infringed	31-Jul-07	X ZR 46/09	aff: invalid	22-Mar-12	EP0852359	20-Dec-96	Computer technology
REV	2 Ni 30/07 (EU)	11-Jun-07		invalid	13-Nov-08						
INF	DU 4A O 136/07	22-May-07	500,000	partly infringed	05-Jun-08	1-2 U 67/08	settlement	10-Nov-11	EP0355391	18-Jul-89	Medical technology
REV	4 Ni 65/07 (EU)	29-Oct-07		invalid	01-Dec-09	n/a	withdrawn	30-Nov-10			
INF	DU 4A O 158/07	14-Jul-07	400,000	infringed	29-Apr-08	1-2 U 47/08	partly infringed	2-Jul-09	EP0885676	13-May-98	Machine tools
REV	4 Ni 14/08 (EU)	07-Mar-08		partly invalid	13-Oct-09	n/a	withdrawn	08-Jun-12			
INF	DU 4B O 284/07	27-Nov-07	1,000,000	infringed	27-Nov-08	1-2 U 2/09	settlement	7-May-09	EP0825350	07-Aug-97	Mechanical elements
REV	10 Ni 6/08 (EU)	08-Oct-08		partly invalid	17-Dec-09						
INF	DU 4B O 310/07	13-Dec-07	500,000	partly infringed	05-Mar-09	1-2 U 44/09	settlement	4-May-11	EP0835737	08-Oct-97	Other special machines
REV	4 Ni 80/08 (EU)	01-Aug-08		invalid	28-Sep-10						
INF	DU 4A O 152/08	29-Jan-08	750,000	infringed	07-May-09	1-2 U 71/09	infringed	n/a	EP0500590	10-Oct-90	Surface technology
REV	3 Ni 62/08 (EU)	16-Oct-08		partly invalid	05-May-10						
INF	DU 4A O 270/08	01-Nov-08	250,000	partly infringed	22-Dec-09	1-2 U 18/10	settlement	1-Sep-11	DE19727527	30-Jun-97	Measurement
REV	5 Ni 123/09	06-May-09		partly invalid	03-Feb-11						
INF	DU 4B O 155/08	25-Jun-08	500,000	infringed	14-Jul-09	1-2 U 97/09	settlement	16-Dec-10	DE19500529	11-Jan-95	Medical technology
REV	4 Ni 40/08	20-Mar-08		invalid	16-Mar-10	n/a	withdrawn	10-Dec-10			

Notes: INF: infringement proceeding. REV: revocation proceeding.

3.7.3   Model Extensions

This appendix considers various extensions of the theoretical model presented in Section 3.3 to show that its predictions are robust to relaxing various assumptions.

Asymmetric information

In the first extension, we assume the alleged infringer knows his litigation costs C_D and c_D, the patent holder, however, only knows the cumulative distribution functions of these litigation costs $F(\cdot)$ and $G(\cdot)$. Let $\bar{C}_D$ and $\bar{c}_D$ denote the average litigation costs.

It is easily seen that the alleged infringer's decision in stage 2 remains unchanged with respect to the baseline model, leaving the model's predictions unchanged as well.

The only change with respect to the baseline model occurs in Stage 1: the patent holder cannot make his decision to file for infringement conditional on the level of the litigation cost, c_D, the infringer has to bear if she decides to challenge the patent's validity. However, he anticipates that the alleged infringer will challenge the patent's validity if and only if $c_D \leq \bar{c}_D$, which happens with probability $G(\bar{c}_D)$. Therefore, the patent holder will file for infringement if and only if

$$G(\bar{c}_D)[\theta\alpha B_P + \theta(1-\alpha)(b_P - L_P) - (1-\theta)(1-\alpha)L_P - C_P] + (1 - G(\bar{c}_D))(\theta B_P - C_P) \geq 0,$$

which can be rewritten as

$$C_P \leq G(\bar{c}_D)[\theta\alpha B_P + \theta(1-\alpha)(b_P - L_P) - (1-\theta)(1-\alpha)L_P] + (1 - G(\bar{c}_D))\theta B_P.$$

Note that the expression on the right hand side is greater than $\tilde{C}_P^{nb}$, which implies that the result that there will be more infringement proceedings under the bifurcated system still holds in this extension.

A qualitatively similar analysis holds if we assume that the alleged infringer has private information about the probability of infringement θ instead of, or in combination with, private information about his litigation costs.

Endogenous infringement

In this extension we add to the baseline game a Stage 0 in which the alleged infringer decides whether to take an action (e.g., use a technology or market an improved version of an existing product) that may be detected by the patent holder and considered as a possible infringement. If this is the case, the game proceeds to Stage 1. Otherwise, the game stops. Let B_D denote the benefit that taking this action brings to the alleged infringer and β denote the probability of detection.

Besides showing that our predictions still hold, we will derive some additional insights into how bifurcation affects a firm's incentives to take actions that can trigger a patent dispute.

We first consider the non-bifurcated legal system as a benchmark.

If $C_p > \tilde{C}_p^{nb}$, then the alleged infringer expects her action to trigger no infringement action even if detected and, therefore, will always take it. However, if $C_p \leq \tilde{C}_p^{nb}$, then she anticipates that a patent dispute will be triggered if the action is detected and she will therefore take the action if and only if

$$B_D - \beta C_D - \theta \alpha \beta L_D \geq 0,$$

which can be rewritten as

$$C_D \leq \frac{B_D}{\beta} - \theta \alpha L_D \equiv C_D^{nb}.$$

In contrast, consider the bifurcated system.

It is straightforward that the alleged infringer's behavior in Stage 2 (i.e., after the patent holder has decided to file for infringement) remains the same as in the baseline model. Therefore, our predictions still hold.

Let us now examine the incentives to take an action that possibly infringes the patent.

If $C_p > \tilde{C}_p^{b}$, then the alleged infringer expects the action to trigger no infringement action even if detected and, therefore, will always take it.

Assume now that $C_p \leq \tilde{C}_p^{b}$, so that the alleged infringer expects her action to trigger a patent dispute if detected.

If $c_D > \tilde{c}_D$, then the alleged infringer also knows that she will not challenge the patent's validity if the action is detected. Therefore, she takes the action in Stage 0 if and only if

$$-\theta \beta L_D - \beta C_D + B_D \geq 0,$$

which is the same as

$$C_D \leq \frac{B_D}{\beta} - \theta L_D.$$

However, if $c_D \leq \tilde{c}_D$, then the alleged infringer knows she will challenge the patent's validity if the action is detected. Therefore, the action is taken in Stage 0 if and only if

$$B_D - \beta \left(C_D + c_D \right) - \beta \theta \left(\alpha L_D + (1-\alpha) l_D \right) \geq 0,$$

which is equivalent to

$$C_D \leq \frac{B_D}{\beta} - \theta \left(\alpha L_D + (1-\alpha) l_D \right) - c_D.$$

Given that

$$\tilde{C}_D^b \equiv \begin{cases} \frac{B_D}{\beta} - \theta \left(\alpha L_D + (1-\alpha) l_D \right) - c_D & \text{if} \quad c_D \leq \tilde{c}_D \\ \frac{B_D}{\beta} - \theta L_D & \text{if} \quad c_D > \tilde{c}_D, \end{cases}$$

the alleged infringer takes an action that will trigger an infringement action (if detected) if and only if

$$C_D \leq \tilde{C}_D^b.$$

It is easily checked that, for both $c_D \leq \tilde{c}_D$ and $c_D > \tilde{c}_D$, it holds that

$$\tilde{C}_D^b < \tilde{C}_D^{nb},$$

which implies that the potential infringer is less likely to (knowingly) infringe a patent under the bifurcated system compared to the non-bifurcated system. This is true for low litigation costs (i.e., $c_D \leq \tilde{c}_D$) because then the alleged infringer incurs the additional cost c_D and faces the risk of interim losses l_D if the patent is infringed but subsequently invalidated. It is also true for high litigation costs (i.e., $c_D > \tilde{c}_D$), because then the alleged infringer prefers not to challenge the patent's validity, which increases the probability that she incurs losses L_D. Note that this result means that firms are less likely to undertake any potentially infringing actions at all. This includes willful infringement, but the empirically more relevant scenario is inadvertent infringement through the use of a technology or improvements over existing technologies/products. Hence, the net effect on innovation is clearly ambiguous.

Benefits of removing uncertainty about patent validity

In this extension, we assume that in cases where the patent is found valid but not infringed, the defendant is unaffected but the patent holder derives a benefit $b_p^{val} \geq 0$ from this decision (e.g., because of an increase in the deterrence of future infringements).

The alleged infringer's behavior is unaffected by this assumption, which implies that our predictions still hold. However, the patent holder's incentives to file for infringement will be higher, because now a challenge to the patent's validity benefits the patent holder with some probability. More specifically, whenever the patent holder expects an infringement proceeding to trigger an invalidity action, i.e., whenever $c_D \leq \tilde{c}_D$, he will file for infringement against the alleged infringer if and only if

$$\theta \alpha B_p + \theta (1-\alpha)(b_p - L_p) - (1-\theta)(1-\alpha) L_p + (1-\theta) \alpha b_p^{val} - C_p \geq 0,$$

which can be rewritten as

$$C_p \leq \theta \alpha B_p + \theta (1-\alpha)(b_p - L_p) - (1-\theta)(1-\alpha) L_p + (1-\theta) \alpha b_p^{val}.$$

Since the right hand side of the equation is greater than $\tilde{C}_p^b$, the patent holder will indeed find it optimal to file an infringement action for a larger parameter space.

Chapter 4

The Timing of Patent Transfers in Europe

4.1 Introduction

One fundamental characteristic of patents is transferability. While the rights to exclude others from making, using, or selling the protected technology originate with the inventor, they can be transferred to anyone who gains ownership of the patent. As tradable property rights, patents facilitate technology transfer and promote the vertical and horizontal disintegration of knowledge-based industries (Hall and Ziedonis, 2001; Arora *et al.*, 2004). Corporations in particular can optimize their innovation and appropriation activities by transferring patent rights via the market for technology (Arora and Gambardella, 2010).[95] The trade in patents is therefore largely seen to promote specialization and to rectify initial mismatch between locus of idea and best-possible utilization, contributing to economic growth (Akcigit *et al.*, 2013; Spulber, 2015).

Nonetheless, patent transfers have long been overshadowed by licensing as subject for research and as empirical evidence of the market for technology (cf. Gambardella *et al.*, 2007; Arora and Gambardella, 2010).[96] More recently, however, studies on patent transfers based on reassignment data for U.S. patents (e.g., Serrano, 2010; Galasso *et al.*, 2013; Serrano, 2013) have spurred great interest among scholars. Unfortunately, no processed data on contemporary patent transfers in other jurisdictions are yet available. Thus, the empirical analysis of patent transfers in Europe poses a particular data challenge due to the fragmented nature of the European patent system. Although the grant process for European bundle patents (*EP*)

---

[95]*Market for technology* is one of a variety of related terms, including *market for ideas*, *market for inventions*, *market for patents*, and *market for innovative control*. For future reference, I consider the *market for patents*, where the ownership of patent rights is transferred between independent entities, to be a subset of the *market for technology*, which also encompasses licensing activities.

[96]The license and transferal of patents differ in several respects. Licensing usually comes with the right to use the technology only, whereas the transfer of a patent conveys full control rights; i.e., to use, license, enforce, and even retransfer the patent.

is centralized, administration after grant is handled by each national patent office separately. Hence, collecting and combining data from multiple authorities is necessary to construct the entire chain of ownership of *EP* patents.

In this study I introduce a novel dataset covering patent transfers in Europe, the *MPI-IC Patent Transfers Data 2015* (hereafter MPIIC-PT2015). Drawing on information from two administrative sources, the German Federal Patent and Trademark Office (*Deutsches Patent- und Markenamt* – DPMA) and the European Patent Office (EPO), the MPIIC-PT2015 contains all registered changes in patent ownership information of domestic patents (*DE*) and patents with Germany as designated country (*EP*) between January 1981 and August 2013. For each *DE* and *EP* patent associated with a change in patent ownership information, I collected information on the identity of current and all prior rights holders, as well as the date of change. For *DE* patents and for *EP* patents which have entered the national phase, the DPMA is a comprehensive data source. To reconstruct the ownership history for *EP* patents prior to the national phase, I complement the DPMA data with changes in patent ownership information previously registered at the EPO from PATSTAT and the European Patent Register.

The MPIIC-PT2015 dataset is unique in its scale and quality. As no single patent authority registers procedural information on all *EP* bundle patents from filing to expiration, reconstructing the entire chain of ownership poses a considerable challenge. Once granted, *EP* patents are split into national parts for each member state of the European Patent Convention (EPC) where the rights holders seek patent protection. To validate the national part of the patent at the respective national patent office, rights holders are subject to validation fees and translation costs. Hence, most *EP* patents are only validated in a subset of EPC countries. Furthermore, rights holders often decide to let some or all national parts lapse prematurely due to the annual burden of renewal fees. Among all EPC countries, Germany has both the highest validation rate (Harhoff *et al.*, 2009) and the highest renewal rate (Van Pottelsberghe de la Potterie and van Zeebroeck, 2008). The DPMA therefore administrates *EP* patents in larger numbers and on average for a longer duration than any other national patent office. From this it follows that the data are probably the most comprehensive records on ownership changes of *EP* bundle patents.

A further challenge in the empirical analysis of patent transfers is their identification and classification within the entire population of registered changes in ownership information. Above all, patent transfers are embedded in very distinct transaction contexts, of which only a subset qualifies as evidence for the market for technology. To define the context of each transaction, I use concepts originating from the institutional economics literature; in particular, the governance structure of the transaction (cf. Williamson, 1985, 1991). Patent transfers may occur between two dependent parties as so-called *hierarchical transactions*. Alternatively, patent transfers can represent technology deals between independent entities; i.e., as *arm's length transactions*. Likewise, patent transfers can be either disembodied or part of what I term *(dis-)integrative transactions*. Patents are often transferred through acquisitions (or spinoffs) of whole companies or business divisions. In these (dis-)integrative transactions, the transfer of codified knowledge goes hand in hand with the transfer of complementary artifacts (i.e., tangible assets) and/or tacit knowledge (i.e., employees).

Taking the governance structure of the transaction into account when analyzing patent transfers is vital, because each transfer type follows its very own economic rationale. Hierarchical transactions are predominantly seen as strategic resource allocations by fiat of the controlling party and thus barely correspond to economic exchanges. Likewise, (dis-)integrative transactions may well be technology-driven, in the form of technology sourcing (e.g., Veugelers and Cassiman, 1999; Grimpe and Hussinger, 2008; Bena and Li, 2014) or corporate venturing (e.g., Parhankangas and Arenius, 2003); however, as long as further assets are involved, it remains unclear whether the gain of patent ownership is the main objective or merely a consequence. All in all, hierarchical and (dis-)integrative transactions certainly prove the transferability of intellectual property rights, but not necessarily the functioning of the market for patents. It thus appears essential – and in line with the notion of market exchanges (cf. Conti *et al.*, 2013) – to focus on arm's length disembodied transactions when using the market for patents as evidence for the market for technology.

In addition to relational distance, the MPIIC-PT2015 incorporates a further fundamental aspect of market analysis: the spatial distance between the transacting parties. While spatial distance plays a key role in the transfer of tacit knowledge,[97] the market for patents does not appear completely detached from geography either (cf. Monk, 2009). Furthermore, since transfers frequently occur at a national or even international scale, patents become subject to tax regime changes with potential implications for their (net) value. Location-specific differences in patent value may have a bearing on the incentives to reallocate patents within a corporate group as well as on the likelihood of market transactions. By using the address information of current and prior rights holders, I define the transaction as either *local*, *national*, or *international*.

As one of its main areas of interest, the literature on technology transfers has identified and extensively studied other entities than corporations as a technology source. These entities include, but are not limited to, public and private research institutions and universities (e.g., Bozeman, 2000). In fact, patents are often held and transferred by entities which are active in

---

[97] For a literature review on different aspects of technology and knowledge transfer, see Battistella *et al.* (2015).

the market for patents but do not operate downstream in the product market (Cotropia *et al.*, 2014). I account for the type of current and prior rights holders with special emphasis on their vertical position and presumed role in the market for patents and distinguish between *individuals (and inventors), corporations (and patent assertion entities), universities, research institutions* and *governmental organizations*.

Accounting for the relational and spatial dimension of the transaction and distinguishing between different types of rights holders, I determine the type of all observed changes in ownership information according to a newly developed taxonomy of patent transfers.

The motivation for assembling the MPIIC-PT2015 is twofold. Patent transfers are of great academic interest on their own, but they may also represent a long disregarded source of distortion in empirical patent studies.

These data will most clearly support the empirical analyses of the market for patents in Europe. While licensing activities and anecdotal evidence of European patent assertion entities that have acquired entire patent portfolios[98] suggest that patent rights are sold and licensed between independent organizations, I am not aware of any study that has quantified the European market for patents on a transaction level. Using a set of arm's length patent transactions, I can examine the activities in the market for patents in Europe. Initial results show an absolute increase in patent transfers since the end of the 1990s, whereas the transfer rate relative to granted patents has remained fairly constant, at about 7 to 8%.

Scholars have also pointed to information asymmetries in patent quality and value as challenges to the functioning of the market for technology (Gambardella *et al.*, 2007; Gans *et al.*, 2008; Palermo *et al.*, 2015). I follow this line of research and investigate how uncertainty over the grant of the patent affects market transactions. This is of particular interest here, since the European patent system shows considerable differences in grant lag and grant rate to the U.S. patent system, which has been the focus of prior studies. The results show a significant effect of the grant event on the hazard rate of the patent being transferred. This suggests that transactions are more likely to succeed when uncertainty over patent validity is resolved.

Besides its direct applicability to studies on the market for patents, the MPIIC-PT2015 dataset also provides a methodological contribution. For a long time, dynamics in patent ownership have only received tenuous treatment in empirical studies using patent data. Several attempts to aggregate patent ownership at business group level have been made (Hall *et al.*, 2001; Thoma *et al.*, 2010; Callaert *et al.*, 2011); however, these datasets take a static view by specifying patent ownership status only at the time of grant or the time of the latest patent

---

[98]For instance, IPCom's acquisition of the mobile telephony patent portfolio developed by Robert Bosch GmbH attracted public attention (cf. the press release of the European Commission, available at http://europa.eu/rapid/press-release_MEMO-09-549_en.htm [accessed: 22 July 2015]).

publication.[99] Empirical studies using conventional patent data are thus prone to bias caused by time-related distortions (Lerner and Seru, 2015). The MPIIC-PT2015 dataset, in contrast, observes all registered ownership changes of any particular *DE* or *EP* patent. This unique panel structure allows to identify the rights holder at a given point in time and eases the construction of time-variant patent portfolios.

The chapter is structured as follows. Section 4.2 provides a short survey of currently available datasets and studies on patent transfers. Section 4.3 introduces a new taxonomy of patent transfers. I then illustrate the collection procedure and scope of the dataset in Section 4.4. Section 4.5 describes the methodology of transfer type classification. Selected descriptive statistics of the data are illustrated in Section 4.6. In Section 4.7, I look closer at the effect of uncertainty on the market for patents. Section 4.8 contains the conclusions.

4.2 Prior Studies on Patent Transfers

Several national patent offices have been systematically registering patent ownership changes since as early as the 19th century.[100] Studies on patent transfers focus either on historical data, covering the supposed heyday of patent transfers during the late 19th century, or on more recent data from the last 30 years.[101]

Historical datasets

Lamoreaux and Sokoloff (1999a,b) analyze records on patent ownership changes in the U.S. in selected periods between 1870 and 1911. They infer that patents were frequently traded despite often-cited information and contracting problems. In follow-on research (Lamoreaux and Sokoloff, 2001; Khan and Sokoloff, 2004), they further find that the flourishing market for technology allowed innovative labor and specialization to be divided among independent inventors. Likewise, Burhop (2010) and Burhop and Wolf (2013) studied ownership changes in Germany between 1884 and 1913. Burhop (2010) finds a general increase in patent trade, even though the transfer rate is substantially lower than in the U.S. during the same period. Observing that heavily cited patents are traded more often, he argues against market failure by information asymmetry between seller and buyer. In the latter study, Burhop and Wolf

---

[99]This can be partly attributed to the fact that, until recently, the main raw patent data provider, PATSTAT, only provided rights holder information based on the latest patent publication.

[100]This is in marked contrast to the rather burgeoning academic interest in patent transfers.

[101]I note there are also several studies using small scale datasets that rely on manually identified patent acquisitions of particular actors in the market for technology (Fischer and Henkel, 2012) or on patent auctions (Sneed and Johnson, 2009; Fischer and Leidinger, 2014; Odasso *et al.*, 2015).

(2013) focus on the geographical dimension of patent transfers and find distance and borders have negative effects on patent transfer. Nicholas and Shimizu (2013) present the numbers for patents involved in sale transactions between 1886, right after the establishment of the Japanese patent system, and 1926. They find a transfer rate comparable to Lamoreaux and Sokoloff (1999a). Andersson (2014) studies ownership changes of Swedish patents between 1871 and 1914 and finds that a legislative change towards a stronger patent regime led to an increase in patent transfers.

Contemporary datasets

Contemporary datasets can be distinguished by the initial structure of the raw data provided by the respective patent authorities. The USPTO offers transfer data in the form of reassignment requests. The raw data here include all requests for reassignment with information on assignor, assignee, and the affected patents. Most other patent offices, such as INPI, EPO, DPMA, and UKIPO, do not disclose reassignment requests in a structured form. Instead, they update ownership information changes in their respective registers, thus providing longitudinal ownership information directly on the patent level. Transaction agreements concerning multiple patents can then be inferred by collapsing the data on the date of change and seller-buyer pairs.

Serrano (2010) provides the first ever large-scale analysis of recent patent transfers in the U.S. (1980-2002), based on reassignment data. He finds that the transfer rate differs across technologies and sectors, as well as by patent age and value. There is considerable follow-on research based on the same data, exploring the implications of the evidently functioning patent market on litigation (Galasso et al., 2013), inventor surplus (Serrano, 2013) and venture lending (Hochberg et al., 2015). Chesbrough (2006), citing early insights from Serrano's then work in progress, provides descriptive evidence for the market for patents, based on U.S. and Japanese data. Analyzing a small subsample in detail, he provides examples of the different types of patent ownership changes the USPTO reassignment data entail. In the same vein, the USPTO recently assembled a patent reassignment dataset of its own, containing all records on registered patent reassignments between 1970 and 2014, which it intends to make publicly available to researchers (Marco et al., 2015). Drivas and Economidou (2015) use a preliminary version of this dataset to look at patent transfers between U.S. states.

Since its September 2010 version, the legal status database in PATSTAT has provided information on patent ownership information changes registered at more than 20 patent offices. These include, most notably, registrations at the EPO and several patent offices in EPC countries. Even though the legal status database initially seems to be a convenient way to capture ownership changes on a patent level, the data remain problematic – even in more recent versions – due to inconsistent legal event codification, incomplete coverage, and ownership information fields that are truncated or completely missing (Martínez, 2011). I am aware of only one project which draws on this data source: Ménière and Dechezleprêtre (2012) present first results from their dataset on the basis of ownership changes for patents granted (*FR*) or validated in France (*EP*) between 1997 and 2009.

In 2013, the EPO changed its internal database and started offering data on applicants and inventors directly from the European Patent Register, holding both current and historical ownership information. Bösenberg and Egger (2014) exploit this new data source to study tax-induced cross-country changes in *EP* patent application ownership between 1996 and 2012.

Table 4.1: Comparison of available historical and contemporary patent transfer datasets

Study	Patent office	Time	Patents	Transfer rate
Historical data				
Lamoreaux and Sokoloff (1999a)	USPTO	1870-71, 1890-91, 1910-11	4,600	14.9%
Burhop (2010)	DPMA	1877-1887, 1889-1913	18,135	8.3%
Nicholas and Shimizu (2013)	JPO	1886-1926	5,000	14.4%
Andersson (2014)	PRV	1871-1914	1,300	7.0%
Contemporary data				
Chesbrough (2006)	JPO	1997-2005	64,352	2.5%
Serrano (2010)	USPTO	1983-2001	170,470	13.5%
Ménière and Dechezleprêtre (2012)	EPO/INPI	1997-2009	56,060	7.0%
Bösenberg and Egger (2014)	EPO	1996-2012	n/a	n/a
Marco *et al.* (2015)	USPTO	1970-2014	n/a	∼ 15.0%

Notes: Transfer rates on the basis of granted patents; DPMA: German Patent and Trademark Office (respectively, German Imperial Patent Office), EPO: European Patent Office, INPI: National Institute of Industrial Property (France), JPO: Japanese Patent Office, RPV: Royal Swedish Patent Office, USPTO: United States Patent and Trademark Office

Judging from the studies above, the propensities of historic and current U.S. patent transfers appear surprisingly similar (see Table 4.1). Conversely, historical studies for European countries cite considerably lower transfer rates. This discrepancy seems to persist even today.[102] Due to varying definitions and data properties across studies and jurisdictions, the extent to which these transfer rates are comparable remains unclear.

---

[102] In the same vein, results from survey data on patenting firms' licensing activities suggest the European market for technology is still relatively underdeveloped (cf. Gambardella *et al.*, 2007).

4.3   Patent Transfer Taxonomy

In its broadest sense, the term *patent transfer* refers to all changes in patent ownership and/or location. Patent transfers include transactions, which involve a change in ownership, as well as patent reallocations, where the rights holder remains the same but the patent is transferred between different locations. To ease the empirical analysis of these and future patent transfer data, I propose a simple taxonomy of patent transfers that takes into account the relational and spatial distance between current and prior rights holders.

4.3.1   Relational and Spatial Distance

Relational distance

The first dimension in my taxonomy captures the relational distance between current and prior patent holders. Determining the relational distance allows for a better interpretation of the economic terms and rationale behind the patent transaction. To define the relational context of a patent transaction, I lean on the concept of governance structure from the institutional economics literature (Williamson, 1985, 1991). I choose a simple dichotomy of relational distance that understands the transacting parties in either a *hierarchical* or *arm's length* relationship.[103] If current and prior rights holders are dependent entities before and after the transaction, I label their relational distance as hierarchical. This applies to patent transactions where the transfer of ownership is embedded in hierarchical governance structure, such as intra-organizational or intra-group transfers. I therefore define organizational dependency as the control one party exerts on the other through direct or indirect ownership. Here, the patent transactions are coordinated by fiat of the controlling party. For instance, a hierarchical transaction occurs when a corporation decides to transfer a patent held by one of its operating subsidiaries to the centralized IP division.

Conversely, I define the relational distance as *arm's length* if the transaction takes place in a market governance structure; that is, the transaction is coordinated by price. Here, current and prior rights holders are organizationally independent parties; the transaction is not affected by

---

[103]I decide to neglect hybrid structures, even though corporations often engage in inter-organizational dealings to gain a competitive advantage (Powell *et al.*, 1996). The technology sourcing literature specifies a variety of hybrid governance structures, such as R&D collaborations, technology alliances and networks, where knowledge transfer is found to be more effective compared to market structures. However, I note that patents as codified knowledge can be easily transmitted across organizational boundaries. Furthermore, trust-based governance structures are primarily chosen to counter opportunism (Dyer and Singh, 1998). Jensen *et al.* (2015) find that patents are effective substitutes for trust in the market for technology. Survey data presented by Caviggioli and Ughetto (2013) also suggests that patents are sold to unaffiliated parties.

hierarchical order.[104] For instance, an arm's length patent transaction occurs through a private sale agreement or a successful public auction.

Besides the governance structure, I also draw a distinction between whether or not the patent transaction involves additional assets. If the patent ownership change takes place in the context of an investment in – or divestment of – complementary artifacts (i.e., tangible assets) or tacit knowledge (i.e., employees), I define the transaction as *(dis-)integrative*. Integrative transactions can therefore refer to the acquisition of a wide spectrum of corporate assets, including product lines, branches, business units, or entire corporations. In the same vein, disintegrative patent transactions refer to corporate spinoffs or buyouts.

The rationale behind treating (dis-)integrative transactions and stand-alone transactions separately is as follows. In the context of (dis-)integrative transactions, the market for technology intersects with the market for firms and/or the market for human capital (Conti *et al.*, 2013). Without further information, I cannot infer which market is driving the transaction – it remains unclear whether the gain of patent ownership is the main objective or merely a consequence.

Considering the above, I categorize the relational dimension between current and prior rights holders as either organizationally equivalent (i.e., *internal*), dependent (i.e., *hierarchical*), part of the process of (dis-)integration (i.e., *(dis-)integrative*), or independent (i.e., *arm's length*). Only the latter category is in line with the notion of exchanges in the market for patents (Conti *et al.*, 2013). This is particularly important in the light of recent research, which finds that a significant share of patent transfers occur inside organizational boundaries (Arora *et al.*, 2011; Caviggioli and Ughetto, 2013). By ignoring this differentiation, patent transfers studies can systematically overstate the activities in the market for patents.

Spatial

The second dimension in my taxonomy refers to the spatial distance between current and prior patent holders. The geography of innovation literature looks at patents and patent citations in a multitude of studies (see Feldman and Kogler (2010) for a literature review) and finds innovation to be spatially concentrated (e.g., Almeida, 1996; Almeida and Kogut, 1997). Even though the transfer of codified knowledge should be less affected by spatial distance, geography apparently still plays a role in the market for patents. Monk (2009), for instance, observes that specialized intermediaries in the market for technology cluster in *Silicon Valley*. Likewise,

---

[104]I note that organizational independence does not necessarily imply that the parties are on equal footing. Asymmetries in power, status or resource-dependence may affect the transaction context as well. The restriction to organizational dependence can be ascribed to feasibility in empirical measurement.

Drivas and Economidou (2015) and Burhop and Wolf (2013) find that spatial distance and borders affect patent flows in the market for patents.

Furthermore, recent research in public economics has studied the effect of tax regimes on patent ownership (e.g., Karkinsky and Riedel, 2012; Griffith *et al.*, 2014). Location-specific differences in tax regimes can affect the frequency and direction of non-local patent transfers due to transfer pricing and arbitrage. Corporations may reallocate patents within their corporate group for the purpose of tax avoidance. Likewise, different (net) patent values across locations can cause unbalanced patent flows in the market for patents.

Figure 4.1: Taxonomy of patent transfers by relational and spatial distance between transacting parties (own illustration)

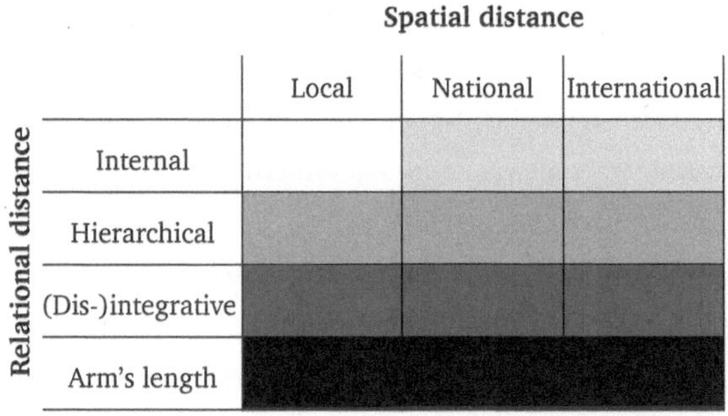

Spatial distance

	Local	National	International
Internal			
Hierarchical			
(Dis-)integrative			
Arm's length			

Relational distance

Notes: ☐ No patent transfer. ▨ Patent transfer but no patent transaction. ▨ Hierarchical patent transaction. ▨ (Dis-)integrative patent transaction. ▨ Arm's length patent transaction.

I define patent transfers as occuring on either a *local*, *national* or *international* scale. If current and prior rights holders are based in the same city or district, the patent transfer stays local. If current and prior rights holders are based in the same country, the patent transfer is national. Finally, if they are from different countries, I classify this as an international patent transfer.

Taking the relational and spatial dimension of patent transfers into account, I can depict the taxonomy as a two-dimensional matrix (see Figure 4.1).

4.3.2   Type of Entity

The literature on technology transfers has identified and extensively studied entities other than corporations as sources for technologies (e.g., Bozeman, 2000). Patents are frequently filed by entities which are active in the market for technology as suppliers, but which do not operate downstream in commercializing the technology (Cotropia *et al.*, 2014). Considering the special interest in the role of current and prior rights holders in the market for patents, I differentiate transacting entities by sector and vertical position; i.e., whether their focus is on invention (upstream) or commercialization (downstream).

As a result, I denote current and prior rights holders as either *individuals (and inventors)*, *corporations (and patent assertion entities), universities, research institutions* or *governmental organizations*. There is no clear classification of the role each type of entity plays in the market for patents. Due to the lack of distinctiveness and varying vertical integration, I cannot order the entities along the innovation value chain.[105] For instance, inventors can engage in entrepreneurial activities and some patent assertion entities may also engage in research and development activities on their own, even though their definition emphasizes the business model of patent acquisition and assertion.[106] With this caveat in mind, I follow Arora and Gambardella (2010), who simply distinguish between *horizontal* transactions between actual or potential rivals and *vertical* transactions between nonrivals, where rivalry is primarily understood as product market competition. By way of illustration, a patent transferred from a university to an operating corporation would be classified as a vertical arm's length patent transaction.

4.4 Data Sources, Structure and Coverage

The MPIIC-PT2015 dataset draws primarily on patent register data retrieved from the German Patent and Trademark Office (*Deutsches Patent- und Markenamt* – DPMA). Along with national patents granted by the DPMA (*DE*), the DPMA is also the authority responsible for registering ownership changes for the national part of European bundle patents (*EP*) after validation in Germany.[107] To construct the entire patent ownership history for *EP* bundle patents, I complement the DPMA data with patent ownership information changes registered at the EPO before validation in Germany. Here, I use the legal events data in PATSTAT to identify changes in patent ownership with detailed information on current and prior rights holders from the European Patent Register for PATSTAT.[108] Figure 4.2 illustrates the main data sources by patent

---

[105]To avoid confusion, I therefore refrain from adding type of entity as a third dimension to the patent transfers taxonomy.

[106]See, for instance, the definition given by the Federal Trade Commission, available at https://www.ftc.gov/policy/federal-register-notices/agency-information-collection-activities-submission-omb-review-49 [accessed: 22 July 2015].

[107]EP patent applications with Germany listed as a designated country enter the DPMA register well before validation, i.e., approximately six weeks after the first publication at the EPO. All legal events associated with *EP* applications registered at the EPO are then forwarded to the DPMA and listed in its register as well. The DPMA register only serves as the main database for collecting information in the national phase of *EP* patents.

[108]I use the PATSTAT version from April 2014.

and phase.[109]

Figure 4.2: Ownership information changes by registering patent authority (own illustration)

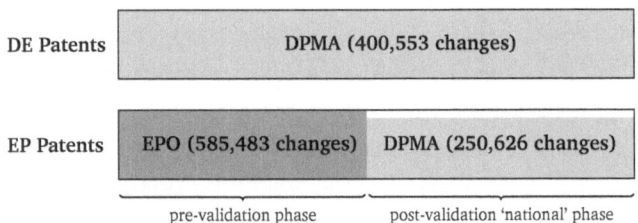

Notes: □ Potential ownership information changes of granted *EP* patents without validation in Germany. *DE* and *EP* patent applications included.

4.4.1 Sources

Ownership changes registered at the DPMA

The DPMA register records bibliographic information, legal status, and procedural events on all published patents and patent applications since 1 January 1981. Using the web-based user interface *DPMAconnect*, I searched for patent applications and granted patents whose procedural records include the change of applicant or patent ownership information. From all *DE* and *EP* patents administered by the DPMA I identified about 505,000 unique patents reporting at least one event of this kind between January 1981 and August 2013. I retrieved the register content for each identified patent as a hierarchical XML file and extracted the following tagged elements into Stata dataset format (**.dta*): the patent application file number as unique identifier, the date of change in patent ownership information, the prior patent ownership information, and the new patent ownership information.

Ownership changes registered at the EPO

For *EP* bundle patents, I complement the DPMA data with ownership information changes registered at the EPO prior to validation in Germany. Here, I draw on the PATSTAT legal events

---

[109]During the first 30 months from priority date of a PCT patent application, the International Bureau at the World Intellectual Property Organization (WIPO) records changes in the person, name, or address of the applicant requested by the rights holder or the receiving patent office. If the PCT application has already entered into the national (DPMA) or regional (EPO) phase, these changes are registered at the patent offices. However, if the PCT application is only filed at the EPO or DPMA afterwards, I do not observe the change. As a result, the original PCT applicant may diverge from the initial rights holder in the data. Information on PCT applicants is available from WIPO. However, without procedural information I am so far unable to specify the date the ownership information was updated.

data to identify the date of change (data available via SQL in file *tls221_INPADOC_PRS*) and on the European Patent Register[110] for the variables containing the prior and updated ownership information (data available in file *reg107_parties*).[111] Overall, I am able to link 471,085 patent applications with a change in ownership information listed in the legal events data with the European Patent Register for PATSTAT.[112]

I retrieved the same information as from the DPMA source; i.e., the patent application file number, the previous and the new patent ownership information, and the date of change in patent ownership information. Since the DPMA register stores the national part of *EP* bundle patents under the EPO patent application file number, I am able to combine EPO and DPMA data with little effort. Due to the superior quality of the EPO data, I only switch to DPMA data after the DPMA officially takes over as administrating authority, i.e., after validation.[113]

4.4.2 Structure and Variables

Structure

I combine the data from both sources and construct a table with information on the latest and all prior patent ownership information in long form with the application authority and number as unique identifier on patent level. In combination with the date of change, this serves as a unique identifier for each change in ownership information.[114] I transform the dataset so that the final MPIIC-PT2015 data file contains a single observation for each registered change in ownership information. Each observation then lists the ownership information of the prior rights holder, referenced as the *seller*, and the ownership information of the new rights holder, referenced as the *buyer*, and the date of change. In case of more than one transfer per patent, the information concerning the buyer in the former observation equals the information concerning the seller in the latter observation.

---

[110]The *EP* register data give details of all newly published applications and patents and recent changes to previously published patents.

[111]PATSTAT's legal events data (*tls221_inpadoc_prs*) also contain legal events for other patent authorities, such as the DPMA. However, the information is of low quality, making the DPMA register the superior data source. For instance, the new owner field (*l509ep* in *tls221_inpadoc_prs*) is truncated to merely 50 characters. In comparison, the corresponding field in the DPMA register has a limit of 232 characters.

[112]Linking these two data sources via the date of change lead to surprisingly few discrepancies. First, for 12,526 (2.7%) of the patent applications, I find no update in the European Patent Register, even though the legal events data suggest a change in ownership information. Here, I use the truncated new owner field (*l509ep*) of the PATSTAT legal events data to fill in information on the new owner. Second, for 836 patent applications (0.2%), the ownership change listed in the legal events data coincides with the first European Patent Register entry. For these cases, I rely on the original application documents to specify the original applicant.

[113]The transition from one responsible authority to the other is not always clear. There is potential for a temporal overlap since the EPO may receive and carry out requests for changes in ownership information if the patent is subject to a pending opposition proceeding.

[114]Multiple changes on the same date are possible. I account for this in the EPO register data by marginally changing the transfer date for coinciding transfers.

Rights holder variables

With separate fields for name, address, and country code, the EPO offers detailed information on the identity and residence of the rights holders. In addition to the name, the DPMA stores address information (country, city, and post codes) in the ownership information field.[115] In contrast to the information retrieved from the European Patent Register, the DPMA falls short only with respect to the rights holder's street address. Unlike the longitudinal ownership information available from other patent authorities in PATSTAT or the USPTO reassignment data, the MPIIC-PT2015 dataset features detailed information on the residence of all current and prior rights holders.[116]

Transfer variables

For the transfer itself, the data include only the registered date of transfer as an original variable. One major disadvantage of the DPMA and EPO ownership change data is the lack of any further classification of the event.[117] I am therefore compelled to identify the nature of the ownership information change on my own (see Section 4.5). In line with this I screen out simple corrections to names and addresses and determine the type of transfer for the remaining subset of observations according to the taxonomy of patent transfers outlined in Section 4.3. To keep the classification methodology as transparent as possible, I state what method and information prompts me to classify the transfer as such. For transactions identified as (dis-)integrative, I present further information on the kind of M&A activity involved (where available).

Patent variables

Each transfer is supplemented with bibliographic and procedural information on the respective patent, obtained from PATSTAT.[118] Added information include dates concerning the filing of the application, the earliest publication, the date of intention to grant, and the official grant date. For *EP* patents I indicate if and when the patent was validated in Germany. Based on information from annual renewal fee payments, I infer the date the patent prematurely expired.

---

[115]The new patent ownership information field is nested, i.e., name, address, and country are stored in separate fields. The prior patent ownership information is stored as one string, but can be easily parsed, as all attributes are consistently delimited.

[116]In those 12,526 cases where I rely on PATSTAT legal events data for the new owner information, I omit information on the patent holder's address and country. I impute these fields according to a conservative method and manually add information where still needed.

[117]The PATSTAT legal events codes for the EPO distinguish between corrections to names and addresses and actual ownership changes, but the classification is neither exhaustive nor reliable. While DPMA and EPO distinguish *de jure* between ownership changes and corrections, this differentiation is not entered into their database.

[118]I use the PATSTAT version from October 2014.

Previous research has found that the likelihood of a patent being transferred depends on a number of patent characteristics (cf. Serrano, 2010). First and foremost, the economic value of patents is positively related to the likelihood of transfer. I therefore include variables such as the number of claims, IPC subclasses, and backward references (patent and non-patent). I include the INPADOC family size to capture patent value and an indicator variable showing whether the patent was filed via the PCT route. I also map the IPC codes assigned to the patents in line with the concordance table developed by the Fraunhofer ISI and the Observatoire des Sciences et des Technologies in cooperation with the French patent office (cf. Schmoch, 2008). The IPC codes are clustered into 34 technology areas, each belonging to one of five main technological areas: (a) electrical engineering, (b) instruments, (c) chemistry, (d) mechanical engineering, (e) other fields.

4.4.3 Coverage and Validity

Coverage

The MPIIC-PT2015 dataset contains all ownership changes for *DE* patents and *EP* patents with Germany as a designated country, except for possible changes for *EP* patents after grant that did not seek validation in Germany. Validation fees and translation costs impede rights holders from validating their patents in all designated EPC countries. In addition, rights holders often decide to let some or all national parts lapse prematurely due to the annual burden of renewal fees. To gain ownership records on the whole population and the maximum duration of any national part of *EP* patents, data from basically all EPC patent offices have to be collected and combined. I refrain from this Herculean task and focus on the German patent office as major source of *EP* patents after validation. Germany has by far the highest validation rate at 95%, followed by France at 80%, and the United Kingdom at 75% (Harhoff *et al.*, 2009, based on data from 2003). Furthermore, Germany also has the highest patent renewal rate among all EPC countries (Van Pottelsberghe de la Potterie and van Zeebroeck, 2008, based on patents filed in 1990). That is, the likelihood of having an *EP* ownership change registered at any national patent office is presumably highest for those at the DPMA.[119]

The MPIIC-PT2015 dataset does not currently contain patents with current or prior co-ownership. This is primarily because the DPMA register only provides one string for all owners, which results in truncated names, missing address information, and indication of country codes

---

[119]As well as the approximately 5% of granted *EP* patents without validation in Germany, the data may also miss transfers of specific national parts of European patents. For instance, the French part of a European bundle patent may be transferred, while the ownership of the German part remains untouched. I believe, however, that fragmented patent transactions, especially in a non-hierarchical context, play a negligible role.

being limited to the first listed owner. Furthermore, the computational effort to link and match multiple prior patent owners to multiple new patent owners is considerably higher. It also requires a different dataset structure and a more complex taxonomy of patent transfers.[120]

Validity

In general, the DPMA and EPO allow for but do not require the recording of ownership changes. The ownership information reflected in the registers is of a declaratory nature, i.e., informative, but not legally binding. Registration is thus not mandatory for and has no effect on the validity of the legal patent transaction. However, registering an ownership change is essential for the subsequent administration and enforcement of the patent. It is only with registration that the new rights holder gains legitimacy to interact with the EPO (or DPMA), the German Federal Patent Court (*BPatG*), which handles validity challenges, and German regional courts (Kraßer, 2009, pp. 468).[121]

In comparison, the costs to register a transfer of rights are negligible. The EPO registers a change in ownership information in the European Patent Register upon the payment of an administrative fee determined by the latest schedule of fees and expenses of the EPO (€ 95 in 2013). The DPMA used to charge fees for registering a transfer of rights, but this has been waived since 2002 (see Table 4.7 in the Appendix for a historical overview of fees).

To maintain the patent, a renewal fee paid by the rights holder is due on an annual basis. Marco *et al.* (2015) finds that U.S. patents which are not renewed are less frequently associated with a prior registered change in ownership, suggesting that changes remain unregistered for (soon-to-be) economically irrelevant patents. I assume this *renewal gap* is less of a problem in my data, since renewal fees at the EPO and DPMA are due at shorter intervals.[122]

Despite the points above, there are widespread concerns that the information held by patent offices does not accurately reflect the actual ownership status of patents. The survey results of a recent report suggest practitioners consider up to 25% of patents worldwide to be lacking accurate and up-to-date ownership information in the patent registers (Oropo, 2015). These doubts in the reliability of patent register information originate primarily from the fact that corporate names are nonstandardized, ultimate ownership can be disguised through shell companies, and ownership changes often go unnoticed. While I am in general accord with the first two

---

[120]Co-owned patents account for about 7% in the raw data. However, on the surface it appears that ownership information changes for co-owned patents mainly take the form of an increase in ownership concentration.

[121]For instance, revocation actions against a particular patent have to be addressed to the rights holder stated in the DPMA register and only he has the right to defend the patent.

[122]The U.S. patent system charges renewal fees only after the 3rd, 7th and 11th year.

points, I believe that coverage of patent ownership dynamics is specific to patent offices.[123]

Even though a large-scale investigation in the validity of the data is beyond my means, I can test whether the DPMA register information captures patent acquisitions that I know have occurred. In order to do so, I draw on the recently revealed patent portfolio of Intellectual Ventures, which contains about 2,000 *DE* and *EP* patents acquired from diverse sources.[124] Of these patents, I am able to unambiguously identify Intellectual Ventures or one of its subsidiaries as the current owner for more than 85% of those patents that are listed in the DPMA register. This result reassures me that the data are in fact a fair representation of patent ownership changes.

4.5 Determining Transfer Type

In this section I first illustrate the challenges associated with determining the type of each patent transfer according to the taxonomy introduced in Section 4.3. I then provide an overview of the determination procedure I followed. Those who like more details are referred to Section 4.9.3 in the Appendix.

4.5.1 Methodological Challenges

The classification of patent transfers in the data poses several challenges. First, as already noted above, some changes in ownership information are not actual ownership transfers. Possible reasons for a change in ownership information changes include administrative corrections of typographical errors, and changes in name, legal form, or address. The initial task is therefore to identify ownership information changes that are not due to an actual patent transfer. These pseudo-changes apply mostly to corporate names, but also to individuals.[125]

Second, even if the change in ownership information reflects an actual ownership transfer between distinct entities, the economic rationale behind it may substantially differ depending on whether the transfer was hierarchical or arm's length. Rule-based methods applied use string similarity measures, based on the reasonable assumption that companies belonging to the same corporate group often share the same name. However, this approach is insufficient. I am therefore required to add additional information on corporate structures to determine the relational context between the transacting parties. To some extent I also infer arm's length patent transactions where the transacting parties are different types of entities, because a majority of cross-sectorial hierarchical relationships appear quite implausible.

---

[123]For instance, Oropo (2015) argues that registration fees deter rights holders from communicating ownership changes. However, the DPMA, among others, charges no fees.

[124]The data are available at http://patents.intven.com/finder [accessed: 22 July 2015].

[125]I observe frequent changes in surname; most likely due to marriage.

Third, the change in patent ownership could simply be a corollary of technology sourcing. To acquire noncodifiable knowledge and expertise, the acquisition of patents occurs in a bundle with further assets, with the whole business unit or firm as the ultimate form. I need to distinguish between disembodied patent transfers and transfers of patents bundled with other assets. The challenge here is that information on patent sales is so scarce that I can only identify these by exclusion. This requires merging additional information on firms' investment and divestiture activities. These data are readily available for M&A activities, but acquisitions of business units and small-scale M&A transactions often remain unregistered.

4.5.2 Determination Procedure

In order to determine the particular context of each patent ownership change in the data, I rely on a mix of rule-based and dictionary-based methods (see Table 4.8 in the Appendix). I use string cleaning and harmonization methods to distinguish changes in ownership from variations in the rights holder's name and legal form. In addition, I perform approximate and partial string matching based on phonetic and probabilistic similarity functions. In this context, one fortunate advantage of the dataset is the fact that I focus on intra-patent comparison of rights holders, where the tradeoff between precision and recall rate (cf. Raffo and Lhuillery, 2009) is much less important.[126] To further identify the relationship between current and prior rights holders, I link corporate rights holders to external business data with information on corporate structures, name changes, and M&A data, with information on mergers and acquisitions (see also Table 4.9 in the Appendix).

Unfortunately, external business data are rarely complete (if at all available) for corporations in the 1980s and early 1990s. Similarly, external data coverage highly correlates with firm size and deal volume. Hence, information on structure and activities of small firms is less likely to be present in external business data. To reduce systematic bias and lack of data coverage, I draw on information from the internet. In particular, I automatically readout internet search engine results of queries containing the names of both transacting parties to gain additional information on their affiliation to each other.

If I cannot determine any affiliation between the transacting parties at the end of the determination procedure (cf. Table 4.2), I consider the transaction as arm's length by default.

---

[126]This is based on the assumption that name similarity is not correlated with the likelihood that two parties meet in the market for technology. Firms active in the same industry and country are more inclined to share similar name tokens, such as industry-specific generic terms ('semiconductors', 'pharmaceuticals') or country-specific legal forms ('GmbH'). Removing legal forms and then generic words alleviates this problem.

Table 4.2: Determination procedure of patent transfer type

Step	Name	Address
Cleaning		
1) Parsing	✓	✓
2) Standardization and disambiguation	✓	✓
3) Match with external identifier	BVDID	NUTS code
- via	OHIM/EPO- & OECD-BVD correspondence tables	OECD Regpat data
Transfer classification		
4) Sector allocation	✓	
5) Seller-buyer pair matching		
- rule-based (Table 4.8)	✓	✓
- dictionary-based (Table 4.9)	✓	

4.6 Descriptive Analysis

In this section I provide a few descriptive aggregate statistics on the results of the patent transfer classification. Before working with a sample of the data that contains arm's length transactions only, I present descriptive statistics of the overall population of patent transfers. For the analysis, I need to eliminate observations with missing entries and inconsistencies in the raw data for the ownership. I further exclude transfers that occurred outside of the time window and those without patent information in PATSTAT. Table 4.10 in the Appendix summarizes patent ownership information changes in the raw data after record cleaning. Ultimately, I observe 1,206,732 changes in patent ownership information in the dataset, which corresponds to 97% of the raw data.

4.6.1 Patent Transfers by Type

I present the absolute number of patent transfers according to the taxonomy in Table 4.3. By far the largest share of ownership information changes concerns transfers where either no transaction occurred (i.e., current and prior rights holder are the same entity) or the transacting parties are in a hierarchical relationship.[127]

Overall, arm's length transactions make up 13.8% of all ownership information changes, while (dis-)integrative transactions follow with at 14%. Most arm's length transaction occur at

---

[127]I refrain from differentiating between internal and hierarchical changes for two reasons. First, it is cumbersome to distinguish between both, since a change in the rights holder's corporate name may include a reorganization of the legal or operational structures of the corporation as well. Second, since the economically relevant rights holder does not change, the distinction seems hardly relevant for subsequent research.

Table 4.3: Distribution of patent transfers by spatial and relational distance

| Relational distance | Spatial distance | | | | | | | |
| | Local | | National | | International | | Total | |
	N	%	N	%	N	%	N	%
Hierarchical	637,209	84.1%	205,177	57.5%	29,173	31.7%	871,559	72.2%
(Dis-)integrative	89,415	11.8%	638,88	17.9%	15,366	16.7%	168,669	14.0%
Arm's length	30,945	4.1%	88,044	24.7%	47,515	51.6%	166,504	13.8%
Total	757,569	100.0%	357,109	100.0%	92,054	100.0%	1,206,732	100.0%

Notes: The sample consists of all changes in ownership information. The unit of observation is at the patent transfer level. Internal and hierarchical transfers combined. Transfers of non-granted patent applications included.

the national level. I present the number of patent transfers by type over time in both absolute and relative terms (see Figure 4.3 below, and Figure 4.9 in the Appendix).

I assume that my classification suffers from historical bias, leaving many transfers as arm's length by default where it is not really the case. I observe from the relative shares of transfer types over time that the quality of this exercise decreases with the age of transfer. Comparing the share of identified (dis-)integrative transactions with arm's length transactions, I see that the relative share of defined arm's length transactions is considerably larger in the period before the late 1990s. This hints at ownership changes in the form of M&A or spinoffs that could not be identified due to missing information on past corporate activities.[128]

The decline in the transfer rate towards the end of the observation period is a consequence of truncation, as some patent applications have not yet been published.[129]

For the remainder of this chapter, I now turn to the most intriguing subset of patent transfers: arm's length transactions. In light of the historical bias in the data, I decide to exclude all patent transfers occurring prior to 1996 to create a less ambiguous set of arm's length patent transactions. The resulting dataset comprises 127,722 observations (I refer again to Table 4.10 in the Appendix for an overview of the different samples in this study).

---

[128]The corporate registers being used (i.e., Amadeus and Orbis) only started assembling data in the late 1990s. Similarly, Zephyr has been covering M&A since 1997, and before 1990 Thomson One only recorded U.S. market deals. It is therefore reasonable to assume that information from the internet search engine readout falls similarly short when it comes to historic corporate activities.

[129]I face right-hand truncation issues due to patent applications that were still unpublished at the time of data collection. I therefore have no information on ownership changes if the patent was published only after August 2013.

Figure 4.3: Patent transfers by relational distance and years

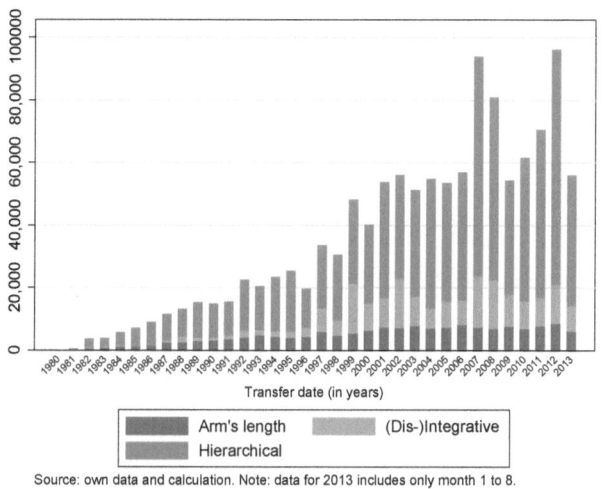

Source: own data and calculation. Note: data for 2013 includes only month 1 to 8.

Notes: The sample consists of all patent transfers. The unit of observation is at the patent transfer level. Transfers of non-granted patent applications included.

4.6.2 Market for Patents

There is strong agreement among scholars that growing technological complexity and conver-
gence has made external technology sourcing increasingly important (Yanagisawa and Guellec,
2009; Arora and Gambardella, 2010). In line with this, the market for patents experienced a
strong increase in activity over the last two decades, with the emergence of new business models, such as patent intermediation (Monk, 2009; Wang, 2010; Hagiu and Yoffe, 2013) and
patent assertion (Chien, 2010). These studies, however, focus exclusively on the U.S. market,
leaving the current status of the market for patents in Europe largely unexplored.[130] While a
detailed analysis of the market for patents is beyond the scope and aims of this study, I am still
able to present initial insights into the absolute and relative frequency of traded patents.

As can be seen in Figure 4.4, in absolute terms the market for patents has experienced an
increase since the beginning of the last decade and appears to have a further upside trend in
the last couple of years. Relative to the number of patents granted, I observe a fairly constant
patent transfer rate of about 7 to 8%.[131]

The rise in the market for patents in the U.S. is partly attributed to the increased patent

---

[130]Although there is some evidence that Europe lags far behind the U.S. in terms of frequency and volume
of patent licensing (cf. Gambardella *et al.*, 2007), I am not aware of any study currently trying to quantify the
European market for patents.

[131]Please note that this transfer rate is not necessarily comparable to the ones stated in Section 4.2. For instance,
Serrano (2010) excludes transfers from individuals in his calculations.

Figure 4.4: Arm's length patent transactions by spatial distance and years

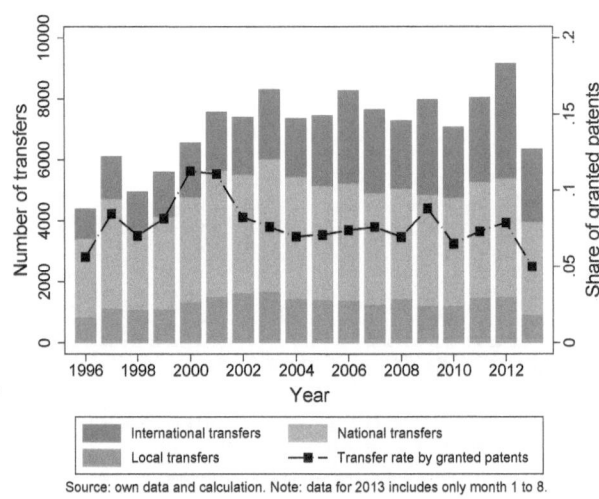

Source: own data and calculation. Note: data for 2013 includes only month 1 to 8.

Notes: The sample consists of arm's length patent transactions. The unit of observation is at the patent transfer level. Transfers of non-granted patent applications included. Data on annual number of granted patents from PATSTAT.

activities of actors on the supply side, such as single inventors, universities and both private and public research institutions. Looking at their activities in the European market for patents, I see a similar pattern as above – a significant increase in vertical patent transfers in the early 2000s (see Figure 4.5). To determine whether the observed patent ownership changes represent technology transfers of European entities or if these transactions mirror the ownership change of U.S. patent equivalents, I distinguish the patent transactions by the origin of the initial patent rights holder.[132] I find that the overall majority of upstream rights holders are in fact European.

To briefly summarize, there is considerable and rising activity in the market for patents in Europe. With inventors, universities and research institutions engaging in the market for patents, I can confirm that vertical transactions between sectors are increasingly common.

4.7   Uncertainty and the Market for Patents

After looking at the overall frequency of patent transactions in the market for patents, I now turn to the impact of uncertainty over the patent's validity on the market for patents.

I follow this line of research and investigate how uncertainty over the grant of the patent

---

[132]The results are similar if I distinguish by the country in which the patent's priority was filed.

Figure 4.5: Arm's length vertical patent transactions by owner origin and years

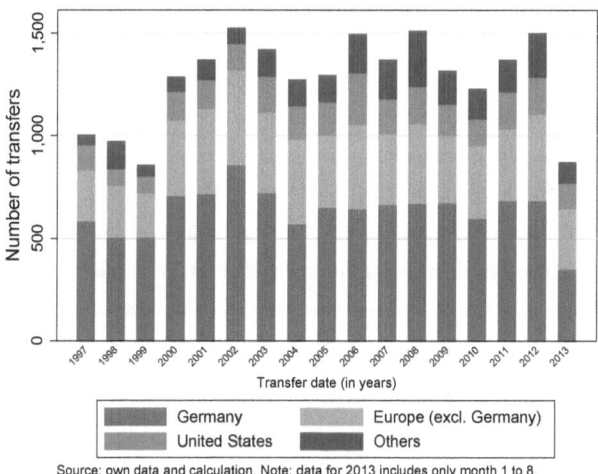

Source: own data and calculation. Note: data for 2013 includes only month 1 to 8.

Notes: The sample consists of arm's length patent transactions with individuals, universities and research institutions as patent seller and corporations as buyer only. The unit of observation is at the patent transfer level. Transfers of non-granted patent applications included.

affects market transactions. This is of particular interest, since the European patent system shows considerable differences in grant lag and grant rate to the U.S. patent system, on which prior studies focused.

Scholars have emphasized information asymmetries in terms of patent quality and value as challenges to the functioning of the market for technology (Gambardella *et al.*, 2007; Gans *et al.*, 2008; Palermo *et al.*, 2015). In fact, Agrawal *et al.* (2015) analyze the process of successful transactions in the market for technology and find bargaining frictions due to uncertainty to be the main source of failure to reach an agreement between seller and buyer. There is good reason to believe that uncertainty also affects negotiations, and therefore completed transactions, in the market for patents.

4.7.1 Effect of Grant on Patent Transfer

One particular source of uncertainty in the patent system is the ex ante unknown outcome of the application process. Neither the applicant nor the potential buyer can foresee in what scope – if at all – the patent will eventually be granted, with considerable implications for the patent's expected value (Gans *et al.*, 2008).

While it must be emphasized that the patents can turn out to be invalid even after grant, the notice of grant is a pivotal event in resolving uncertainty over patent scope and validity that may alleviate negotiations over the terms of the patent transaction. In fact, previous studies have found that the grant event has a positive effect on the timing of contractual licensing agreements (Gans *et al.*, 2008) and research collaborations (Czarnitzki *et al.*, 2015).

However, patent grant does not occur instantly. Particularly at the EPO, the time until grant (i.e., the length of patent examination) can be considerable and varies significantly (Harhoff and Wagner, 2009). This grant lag is affected by several factors at the macro level (capacity constraints at the patent office), patent level (the number of pages and claims), and claim level (controversial and complex claims) (van Zeebroeck *et al.*, 2008; Harhoff and Wagner, 2009).[133] If the outcome of the bargaining process between seller and buyer is subject to frictions due to prolonged uncertainty regarding the patent's validity, grant lag has a direct impact on the market for patents.

Having obtained data on patent ownership changes both before and after patent grant, I study the effect of grant on the event of a successful patent transaction. For this purpose I link the timing of transfer with the moment when the applicant learns about the patent office's intention to grant the patent.

4.7.2 Empirical Model and Results

Empirical model

I analyze the effect of patent grant on the duration outcome of patent transfer, i.e., the point in the patent's life when the (first) transfer occurs. Analogous to Gans *et al.* (2008), I use a Cox proportional hazard rate model with time-varying regressors which is specified as a continuous-time hazard rate function. The model consists of a nonparametric baseline hazard rate and a multiplicative term allowing regressors to have a proportional impact relative to the baseline. I let $h_{TRANSFER}$ denote the hazard rate of the event of transfer stratified by the patent's technology area, l, and including rights holder and patent characteristics, Z, as controlling variables:

$$h_{TRANSFER}\left(t, POST\_GRANT_i^t, l_i, Z_i\right) \tag{4.7.1}$$
$$=h^{l_i}(t)\exp(\beta_0 + \beta_Z Z_i + \beta_{GRANT\_LAG}\ GRANT\_LAG_i +$$
$$+\ \beta_{POST\_GRANT}\ POST\_GRANT_i^t + \epsilon_i),$$

---

[133]It must be noted that applicants can also influence grant delay (Harhoff and Wagner, 2009; Mejer and van Pottelsberghe de la Potterie, 2011).

where the term β_{POST_GRANT} represents the effect of *POST_GRANT* on the *TRANSFER* hazard rate. *POST_GRANT* represents a time-variant dummy variable indicating whether the patent has already been granted at a given time. For identification I use the variation in grant as well as transfer lags across technologies and in the timing of transfer relative to the grant date. The potential bias from a correlation between grant lag and unobserved heterogeneity in the hazard rate, ϵ_i, can be distinguished from the impact of patent grant, since ϵ_i always affects the hazard rate, while the variable *POST_GRANT* only impacts the hazard rate after patent grant.[134]

Table 4.4: Summary statistics

Variables	Mean	Std. dev.	Min	Max
Patent characteristics				
Post-intention to grant (d)	0.65	0.48	0	1
Grant lag (in months)	55.81	25.13	8	187
Post-grant (d)	0.43	0.50	0	1
Patent characteristics				
Backward citations (patents) (ihs)	2.16	0.62	0	6
Backward citations (nonpatent literature) (ihs)	0.45	0.78	0	5
Claims count (ihs)	3.24	0.67	0	6
IPC subclass count (ihs)	1.40	0.50	1	3
Family size (INPADOC) (ihs)	3.83	0.75	1	8
PCT filing (d)	0.53	0.50	0	1
Year of patent application/priority	1998.16	3.99	1991	2008
Patent technology area				
Chemistry (d)	0.30	0.46	0	1
Electrical engineering (d)	0.21	0.41	0	1
Instruments (d)	0.19	0.39	0	1
Mechanical engineering (d)	0.23	0.42	0	1
Other (d)	0.06	0.23	0	1
Patent holder characteristics				
University of research institution (d)	0.06	0.20	0	1
N		16,526		

Notes: Transformation 'ihs' represents the inverse hyperbolic sine of variable, i.e., $\text{ihs}(x) = ln(x + \sqrt{1 + x^2})$.

To capture the characteristics of the patent that may influence the baseline hazard rate of patent transfer, I refer to the standard patent value and scope indicators (see Table 4.4 for a summary of the variables in this analysis). In particular, I include family size, the number of claims, patent and nonpatent backward references, and IPC subclasses.[135] I also account for

---

[134]This is true if *POST_GRANT* is independent of ϵ_i given the stratification and controlling variables.

[135]I omit forward citations as a value indicator due to potential endogeneity with patent transfer.

the entity type of the initial rights holder in a simplified manner by indicating whether the rights holder represents a university or research institution.

I apply the hazard model illustrated above to a sample of identified arm's length patent transactions. I focus on *EP* bundle patents due to a lack of information on the exact date the DPMA communicated the intention to grant to the rights holder. To reduce sample selection bias, I exclude all patents that are not validated in Germany and those withdrawn in the first two years after grant.[136]

Results

I define grant lag as the time between the filing of the patent application and the date the EPO's intention to grant the patent was communicated to the rights holder. Transfer lag represents the time between the filing of the application and the date the transfer was registered at the EPO. In line with Gans *et al.* (2008), I assume the internal communication to the rights holder about the intention to grant the patent is the primary event that reduces uncertainty.[137] I also observe administrative lag, i.e., the time between the intention to grant and the official grant date. As can be seen in Table 4.5, the average duration differs by more than a year between some technology areas.

Table 4.5: Timing of intention to grant, transfer and grant event by main technology area

	Electrical engineering		Instruments		Chemistry		Mechanical engineering		Other fields	
Events	Mean	Std. dev.	Mean	Std. dev.	Mean	Std. dev.	Mean	Std. dev.	Mean	Std. dev.
Intention to grant − application filing	64.00	27.38	62.19	24.22	59.38	24.78	49.69	19.52	49.80	19.01
Transfer − application filing	72.57	35.79	65.69	32.90	64.53	32.08	56.09	29.06	56.23	27.89
Patent grant − intention to grant	6.57	12.00	7.11	12.02	7.38	12.12	5.42	6.83	5.77	8.01

Notes: The sample consists of arm's length transactions of granted *EP* filings with validation in Germany. Duration reported in months. The unit of observation is at the patent transfer level.

Figure 4.6 shows the timing of patent transfers relative to the intention to grant event. I observe a striking rise in patent transfers in the months after the rights holder learns about the successful outcome of the examination process. However, a significant share of patents is transferred considerably before the grant event already.

I now turn to the results of the Cox hazard model presented in Table 4.6. I observe through-out all specifications (T1-T5) that the event of intention to grant has a significant and highly

---

[136]Mejer and van Pottelsberghe de la Potterie (2011) note that about 10% of the patents where the EPO informs the rights holder about the decision to grant the patent do not enter the national phase in any EPC country.

[137]Communication happens in written form; thus the rights holder can provide substantial evidence to third parties.

Figure 4.6: Timing of patent transactions relative to intention to grant event

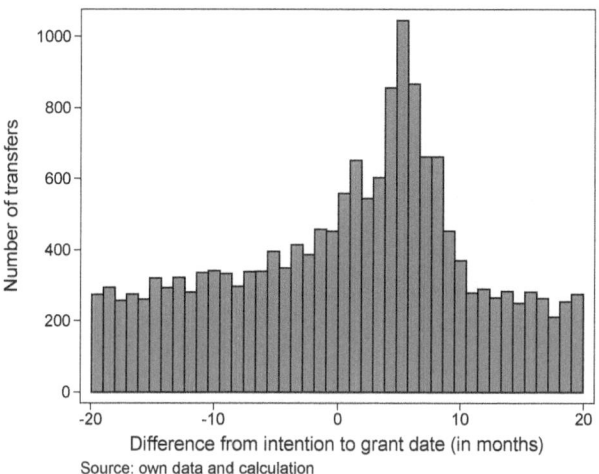

Difference from intention to grant date (in months)
Source: own data and calculation

Notes: The sample consists of arm's length transactions of granted *EP* filings with validation in Germany. The unit of observation is at the patent transfer level. Truncated at ± 20 months.

positive effect on the hazard rate (I also state the respective coefficients to ease interpretation). The same is true if I account for grant lag as well as the subsequent date of official grant – the effect remains robust. The findings indicate that the EPO's communication to grant increases the hazard rate of achieving a patent transaction agreement by more than 40%. Controlling for patent characteristics, I see that some patent value indicators appear to affect the hazard rate. Nonpatent backward citations decrease the hazard rate, which may be due to the fact that citations to scientific articles indicate that the patent is based on science rather than on established technologies (Carpenter *et al.*, 1981). That is, the patent may take longer to become applicable for commercialization. Patents part of a large family and patents entering the EPO via the PCT route have a lower hazard rate. Apparently, more valuable patents remain longer with the initial rights holder. Interestingly, I find that rights holders that do not operate in the product market take shorter to sell their patents.

Table 4.6: Cox proportional hazard model results: incidence of patent transfer

	(T1)		(T2)		(T3)		(T4)		(T5)	
	Hazard ratio	Coefficient	Hazard ratio	Coefficient	Hazard ratio	Coefficient	Hazard ratio	Coefficient	Hazard ratio	Coefficient
Grant process										
Post-intention to grant (d)	1.498***	0.404***	1.589***	0.463***	1.594***	0.466***	1.594***	0.466***	1.550***	0.438***
	(0.033)	(0.022)	(0.048)	(0.030)	(0.048)	(0.030)	(0.048)	(0.030)	(0.047)	(0.030)
Grant lag (in months)			1.001**	0.001**	1.001**	0.001**	1.002***	0.002***	1.002***	0.002***
			(0.000)	(0.000)	(0.000)	(0.000)	(0.000)	(0.000)	(0.000)	(0.001)
Post-grant (d)					1.425***	0.354***	1.389***	0.328***	1.288***	0.253***
					(0.088)	(0.062)	(0.086)	(0.062)	(0.081)	(0.063)
Patent characteristics										
Backward citations (patents) (ihs)							1.007	0.007	0.999	−0.001
							(0.014)	(0.014)	(0.014)	(0.014)
Backward citations (nonpatent literature) (ihs)							0.983	−0.017	0.974*	−0.026*
							(0.012)	(0.012)	(0.012)	(0.012)
Claims count (ihs)							1.017	0.017	0.987	−0.013
							(0.013)	(0.013)	(0.013)	(0.013)
Family size (INPADOC) (ihs)							0.930***	−0.072***	0.949***	−0.052***
							(0.012)	(0.013)	(0.012)	(0.013)
IPC subclass count (ihs)							0.959*	−0.041*	0.980	−0.020
							(0.016)	(0.017)	(0.017)	(0.017)
PCT filing (d)							0.998	−0.002	0.948**	−0.054**
							(0.017)	(0.017)	(0.017)	(0.018)
Patent holder characteristics										
University or research institution									1.279***	0.246***
									(0.055)	(0.043)
Patent holder country	No		No		No		No		Yes***	
Application year effects	No		No		No		No		Yes***	
Observations (unique)	16,526		16,526		16,526		16,526		16,526	

Standard errors in parentheses; * $p < 0.05$, ** $p < 0.01$, *** $p < 0.001$

Notes: The sample is defined as EP bundle patent hazard sample (cf. Table 4.10). The unit of observation is at the transfer level. Standard errors are clustered by rights holder. Baseline stratified by technology area. Column 'Coefficient' represent exponentiated coefficients. Transformation 'ihs' represents the inverse hyperbolic sine of variable, i.e., $ihs(x) = ln(x + \sqrt{1 + x^2})$.

One explanation could be that universities and research institutions prefer to sell the patent as soon as possible due the lacking possibility to commercialize it on their own. Another explanation is that patent transactions between research institutions and corporations are often part of research and development agreements. In this case, the search and bargaining process between seller and buyer could have already started prior to patent filing.

Discussion

Even though the main empirical results appear highly robust across all specifications, there are arguments against direct causality between patent grant and the likelihood of patent transfer. The first issue is that the data capture the date the ownership change is registered at the patent office, not the date of the patent transaction agreement. To strengthen the view of the causal effect on patent transfer, I would need to know whether the legal event that I observe, i.e., the registration of ownership information change, follows directly the actual patent transfer. A divergence of these two dates may originate from either the EPO's slow processing of registration requests or the rights holder's decision to defer the registration until grant.

Figure 4.7: Time lag between request for transfer of right and registration (density)

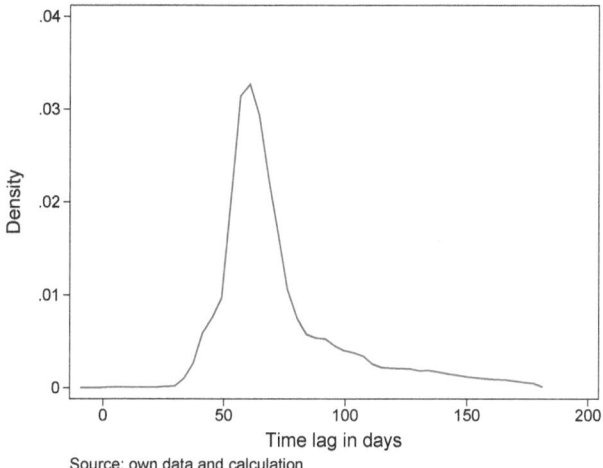

Source: own data and calculation

Notes: The sample consists of arm's length transactions of granted *EP* filings with validation in Germany where registration request date available. The unit of observation is at the patent transfer level. Truncated at 180 days.

To gain insights into the registration process at the EPO, I identified the digital documents for the correspondence between the EPO and the rights holder in the online file inspection system. All documents are listed chronologically with the date of receipt (or dispatch).[138] Looking at the time lag between the request for a change in ownership information and its actual recording by the patent office, I see the usual lag at the EPO is just between 40 and 70 days (cf. Figure 4.7).[139]

Unfortunately, I have no information on the date of market transactions to identify any deferment by the patent holder. An alternative approach is to investigate whether patents in transactions involving multiple patents are registered simultaneously or individually whenever the respective patent is to be granted. From Figure 4.8, I see the difference in registration dates for transactions involving more than one patent is highly skewed, while the difference in grant dates is more equally distributed (in Figure 4.10 in the Appendix I compare the distributions more specifically in a quantile-quantile plot). I consider this a strong signal that ownership registration usually occurs at the time of the actual deal agreement and is not requested whenever grant is due.

Figure 4.8: Time lag between request for transfer of rights and registration (density)

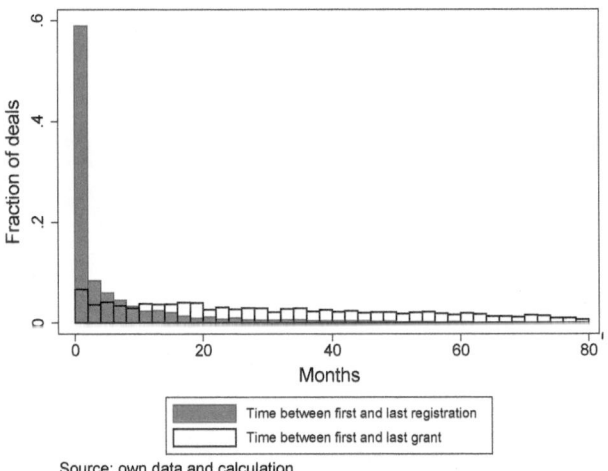

Source: own data and calculation

Notes: The sample consists of arm's length transactions of granted *EP* filings with validation in Germany. The unit of observation is at the deal level. Deals are defined as all patents transferred between the same seller and buyer. Deals with less than two patents excluded. Truncated at 80 months.

It should also be noted that there are benefits from grant delay in the form of cost deferral

---

[138]Outgoing communications become available online on the day after the date of dispatch. Incoming communications become available once the filed document has been noticed and coded by the EPO.

[139]In contrast, Marco *et al.* (2015) observe an average recording lag of more than 6 months at the USPTO.

and prolonged uncertainty for third parties (Harhoff and Wagner, 2009).[140] Patent grant and the subsequent national validation occasion costs in the form of validation, translation, and renewal fees. The consequence may be that applicants induce the grant only after a patent transaction agreement has been reached, implying reverse causality. However, I observe in the data a similarly pronounced increase in patent transactions around the official grant date at the DPMA, even though the grant of *DE* patents prompts no additional costs.[141]

While I cannot exclude that the rights holder may benefit from maintaining uncertainty on the patent's validity when bargaining, I believe this is more likely to apply for patent 'lemons'. Since the analysis focuses on patents that successfully survived the examination process, I am reluctant to see strategic behavior as the driving force behind the identified patent grant effect.

4.8   Conclusion

The MPIIC-PT2015 dataset is the result of extensive data work. Collecting and interpreting data on patent ownership changes in the fragmented European patent system is a resource-intensive task. To determine the particular type of each observed patent transfer, I applied a multitude of established and innovative rule-based and dictionary-based methods. While the classification of patent transfers in early cohorts can still be improved, I believe that the MPIIC-PT2015 dataset will be useful in the empirical analysis of the market for patents as well as other research areas that remained untouched in this study, such as transfer pricing or patent aggregation.

In a first descriptive analysis of the activities in the market for patents in Europe, I found there has been a considerable increase in absolute patent transfers since the late 1990s. The transfer rate by granted patents per year remains relatively constant at 7 to 8% and seems, differences in calculations put aside, significantly lower than transfer rates found for the U.S. market for patents. However, given the scrutiny applied to the identification of arm's length transactions, this discrepancy may partly be a method's artefact. I further observe new actors, in particular universities and research institutions, participating in the market for patents. The increase in vertical patent transactions suggests that patent rights facilitate technology transfer from independent inventors, universities, and research institutions to corporations.

More specifically, I looked at the timing of patent grants and the event of patent transfers. In short, the results reveal that the internal communication of patent grant to the rights holder

---

[140]In a similar vein, Mejer and van Pottelsberghe de la Potterie (2011) argue that grant lag is partly due to the applicants' behavior, who deliberately defer the grant date.

[141]The DPMA does not discriminate between patent applications and granted patents in terms of renewal fees. There are also no validation or translation costs associated with *DE* patents when granted.

significantly increases the hazard of patent transfer. These findings are consistent with the notion a decrease in uncertainty regarding a patent's validity and scope facilitates the success of negotiations between patent sellers and potential buyers. Considering the grant lag at the European Patent Office, this finding indicates some frictions in the market for patents in Europe.

A corollary of this study is the general result that a considerable number of patents are subject to ownership changes. This has implications for research relying on patent data based on rights holder information at the time of official grant or latest publication. In particular, analyses on static patent data may produce a distorted image of the patenting activities of particular entities and sectors. Furthermore, when linking inventors and citations to subsequent instead of to initial rights holders, the empirical analysis of inventor mobility and knowledge spillovers becomes vulnerable to overestimation. The findings of this study hopefully draw attention to the dynamics in patent ownership for an even broader set of innovation literature.

4.9 Appendix to Chapter 4

4.9.1 Figures

Figure 4.9: Relative share of patent transfers by relational distance and years

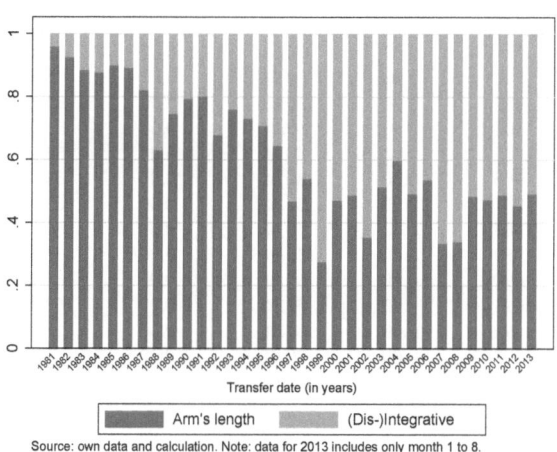

Notes: The sample consists of all arm's length and (dis-)integrative patent transactions. The unit of observation is at the patent transfer level. Transfers of non-granted patent applications included.

Figure 4.10: Time difference between first and last date of registration and grant by deal

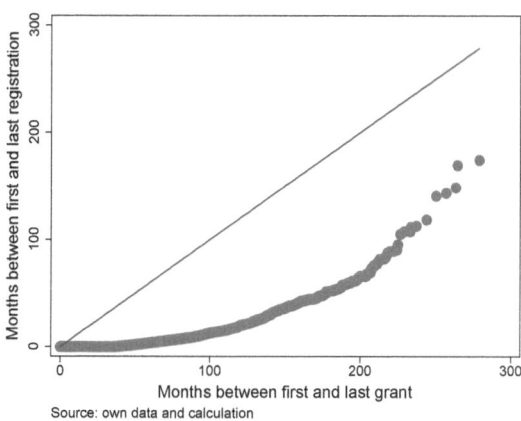

Notes: The sample consists of arm's length transactions of granted *EP* filings with validation in Germany. The unit of observation is at the deal level. Deals are defined as all patents transferred between the same seller and buyer. Deals with less than two patents excluded.

4.9.2   Tables

Table 4.7: Fees for registering changes in patent ownership information by patent office

| | DPMA | | EPO | |
Year	Change in name or address	Transfer of rights	Change in name or address	Transfer of rights
1981	0.00	30.68	0.00	51.13
1982	0.00	30.68	0.00	58.80
1983	0.00	30.68	0.00	58.80
1984	0.00	30.68	0.00	58.80
1985	0.00	30.68	0.00	58.80
1986	0.00	30.68	0.00	63.91
1987	0.00	30.68	0.00	63.91
1988	0.00	30.68	0.00	63.91
1989	0.00	30.68	0.00	63.91
1990	0.00	30.68	0.00	63.91
1991	0.00	30.68	0.00	63.91
1992	0.00	30.68	0.00	76.69
1993	0.00	30.68	0.00	76.69
1994	0.00	30.68	0.00	76.69
1995	0.00	30.68	0.00	76.69
1996	0.00	30.68	0.00	76.69
1997	0.00	30.68	0.00	76.69
1998	0.00	30.68	0.00	76.69
1999	0.00	30.68	0.00	76.69
2000	0.00	35.79	0.00	76.00
2001	0.00	35.79	0.00	76.00
2002	0.00	0.00	0.00	76.00
2003	0.00	0.00	0.00	75.00
2004	0.00	0.00	0.00	75.00
2005	0.00	0.00	0.00	75.00
2006	0.00	0.00	0.00	75.00
2007	0.00	0.00	0.00	80.00
2008	0.00	0.00	0.00	80.00
2009	0.00	0.00	0.00	85.00
2010	0.00	0.00	0.00	85.00
2011	0.00	0.00	0.00	90.00
2012	0.00	0.00	0.00	90.00
2013	0.00	0.00	0.00	95.00

Notes: All fees in euros. Fees stated per patent or patent application. Fees before 2002 (DPMA) and 2000 (EPO) converted from DM to € in accordance with official exchange rate.

Table 4.8: Rule-based methods for the positive identification of transfer type

Matching method	Matched on string			Example	
	Original	Cleaned	Nongeneric	Prior rights holder	Current rights holder
Deterministic string matching					
Same string	✓	✓	✓	Braun GmbH	Braun GmbH
Same tokens	✓	✓	✓	Meyn Machinefabriek B.V.	Machinefabriek Meyn B.V.
Substring	✓	✓	✓	Kassai K.K.	Aprica Kassai K.K.
Same nongeneric first word		✓	✓	Dowa Mining Co., Ltd.	Dowa Holdings Co., Ltd.
Acronym	✓		✓	Asea Brown Boveri AG	ABB (Schweiz) AG
Phonetic string similarity matching					
Soundex		✓	✓	ConocoPhilips Co.	ConocoPhillips Company
Probabilistic string similarity matching					
Levenshtein		✓	✓	Flielinghaus GmbH	Frielinghaus GmbH
Jaro-Winkler		✓	✓	Codivien LP	Covidien LP
Weighted 2-gram		✓	✓	A. Ahlström Corp.	Ahlstrom Machinery Oy

Table 4.9: Data sources used for the dictionary-based positive identification of transfer type

External database	Version date	Link with rights holders via	Positive identification of transfer types					
			Same firm	Same group	M&A	Joint venture	Spinoff	Market transaction
Firm level databases								
Amadeus	2006, 2011	BVDID / string	✓					
Orbis	2014	BVDID / string	✓	✓				
M&A databases								
Thomson	2014	String		✓	✓	✓		
Zephyr	2014	BVDID / string		✓	✓	✓		
Patent office databases								
USPTO reassignments	2014	String	✓		✓			
Internet search engine results								
Bing top 10 results	2015	String	✓	✓	✓	✓	✓	✓

Table 4.10: Overview and definition of subsamples

	Patent Office		
Sample definition	EPO	DPMA	Total
All changes in patent ownership information			
N	585,483	651,179	1,236,662
% of subsample / full sample	47.34	52.66	100.00
– without missing entries			
N	584,015	650,921	1,234,936
% of subsample / full sample	47.23	52.64	99.86
– within time frame			
N	560,149	648,294	1,208,443
% of subsample / full sample	46.35	53.64	97.72
– matched with patent information			
N	559,665	647,041	1,206,706
% of subsample / full sample	46.38	53.62	97.58
Arm's length patent transactions			
N	70,318	96,186	166,504
% of subsample / full sample	42.23	57.77	100.00
– with transfer date after 1995			
N	54,360	73,362	127,722
% of subsample / full sample	47.34	52.66	76.71
EP bundle patents hazard rate sample			
N	9,601	6,925	16,526
% of subsample / full sample	58.10	41.90	100.00

Notes: The unit of observation is the transfer of patent. Missing entries refer to missing and incoherent ownership information. Time frame is January 1981 to August 2013. Patent information refers specifically to information on IPC subclasses and application filing date. EP bundle patents hazard rate sample consists of first time transfers granted EP bundle patent transfers filed after 1990, with no lapse in first two years of national phase and intention to grant date available.

4.9.3   Details on Transfer Type Determination

Sector allocation

The motivation behind allocating all rights holders to their sector is threefold. First, it expands the range of possible research to sector-specific questions. Second, I apply different methods to determine transfer type depending on seller and buyer's sectors. For instance, the strings between individuals are differently matched with each other than those between corporations. Furthermore, in most cross-sectoral patent transactions the relational distance between seller and buyer can be automatically inferred as arm's length.

Unfortunately, there is no codification of entity type available in EPO or DPMA databases that would allow me to single out transactions between rights holders from different sectors. I identify rights holders as either *corporations, individuals, universities*[142], *research institutions*, or *governmental organizations*. I follow the allocation methods applied in Hall *et al.* (2001) and Callaert *et al.* (2011) and search for sector-signaling keywords in the name of the rights holder. I accomplish this by using keyword dictionaries (census first name lists, university lists, NPE lists, etc.). I further differentiate individuals by their relation to the respective patent; i.e., I identify those individuals that are also listed as innovators. Similarly, I identify corporations that are commonly seen to fulfill the characteristics of patent assertion entities.[143] Acquisitions by patent assertion entities can be usually understood as arm's length transactions.

String matching (rule-based methods)

The idea behind deterministic string matching is to directly compare the string variables containing information on the owner and buyer. Conversely, probabilistic matching (cf. Raffo and Lhuillery, 2009) can account for minor variations in string composition. Matching on the basis of strings requires the names of current and prior rights holders to be cleaned and standardized prior to the matching process in order to increase the number of direct matches and the quality of probabilistic matches. There are small, mostly noisy but sometimes even systematic differences in the way names are recorded. These mostly include spelling variations, typographical errors, and optional prefixes and suffixes.[144]

Because the precision rate suffers from standardization, I construct two strings that differ in their degree of cleaning. I harmonize the person name strings of the transacting parties according to established string cleaning methods (cf. primarily Magerman *et al.*, 2009). I first

---

[142]This includes hospitals.

[143]Here, I draw on an expansive list from a commercial IP intelligence provider, available at http://www.rpxcorp.com [accessed: 22 July 2015].

[144]For instance, the DPMA provides details on the incorporation of U.S. firms ('n. d. Ges. d. Staates Delaware'), while the EPO usually omits these.

convert all strings to uppercase, standardize characters to the ISO basic Latin alphabet, harmonize generic words, and remove punctuation marks. I move legal form and subsidiary information as well as content in parentheses to separate variables. In a further step, I remove generic prefixes and suffixes as well as nonessential toponyms to account for domestic subsidiaries of multinational corporations.

For individuals, I found the approach described above unsatisfactory. Ownership information changes of individuals are primarily due to name changes and transfers to family members in the form of inheritance. Here, the above methods fall short, since names (first or last) remain the same, which may lead to false positives if the same probabilistic methods are applied. I therefore pursued the following approach for individual rights holders. I parsed first, last, and middle names and subsequently used the first name to identify the gender of the person by matching with census data. I decided to classify the transfer as follows. If I encountered the same or similar last name but a different first name, I assumed inheritance. If the first name was the same or very similar, the last name was different, and the individual's gender was female, I assumed a name change due to marriage.

The matching approach itself is as follows. I apply a hierarchical approach by always checking the least treated string first. I execute direct matchings based on the original, cleaned, word-ordered, and non-generic string. I identify those cases where one string is a substring of another. I also check whether the first word of the non-generic strings is the same and detect acronyms by comparing capital letters. Probabilistic and phonetic string similarity matchings are applied to cleaned and non-generic strings in the same manner. I check for phonetic similarity and apply probabilistic algorithms, namely Levenshtein distance, Jaro-Winkler distance and n-gram (cf. Raffo and Lhuillery, 2009).

If the strings are found to be equal or sufficiently similar, the patent transfer is interpreted as an internal/hierarchical transaction.

Link to external data (dictionary-based methods)

I match current and prior rights holders with several external dictionaries. These dictionaries are helpful to determine any ongoing or previous affiliation between the transacting parties. Patent data by default have no unique identifier available that allows for a direct match with external firm level data. In lieu of this, prior studies use the names indicated on patent documents and the corporate names contained in firm level databases for simple string matching. I refrain from creating a new link between rights holders and corporate register information and instead draw on already existing correspondence tables created by EPO-OHIM (2013). Using the added *BVDID* identification number, rights holders can then be linked to several corporate registry databases provided by Bureau van Dijke, most notably Amadeus, Orbis, and Zephyr. Because the data captures *DE* patent rights holders that are not part of the correspondence tables, I extend the link via a deterministic match on country and cleaned name variable.

Information on shareholders, subsidiaries, and branches contained in Amadeus and Orbis enables the construction of the corporate structure and the identification of all direct and indirect links between corporations belonging to the same group. These corporate registry data are static; i.e., they report only the current corporate structure. In the case of mergers and acquisitions, I face the problem that firms are treated as part of a group before they actually become part of it. To control for any activities in the market for firms, I draw on M&A databases, in particular Zephyr and Thomson One.[145] I must note that M&A deals may trigger several kinds of patent transactions, because patents are often centrally held by the parent corporation. Hence, patent transfer may go indirectly from vendor to acquirer without the involvement of the target. I therefore control for any potential links between two corporate groups associated with an M&A deal. If the patent had been transferred between two corporations before they became part of the same group, I consider them as arm's length instead of hierarchical.

Internet search

To reduce systematic bias due to sample selection issues in terms of exit of firms and lack of data coverage of available firm level databases, I additionally draw on a novel approach to automatically readout internet search engine results.

I first collapse the data of unclassified patent transfers to unique seller buyer pairs and combine the two cleaned firm name strings into a single search query. I send these search queries to the internet search engine *Bing*.[146] The ten search results presented by Bing on its first search results page are then automatically extracted and stored in a local dataset. For each search result, I have the title of the web page, the link, and most importantly, the 'snippet,' a short excerpt of the text on the website that commonly contains the search keywords in context. I remove misleading and irrelevant search results, such as user content websites, product portals, professional networks, and patent information providers, and I give priority to financial news portals, company registers, and company websites. I then automatically determine the transfer type by searching for keywords signaling the relational distance in title, link, and snippet of the respective search result. All keywords found are then ordered by decreasing ambiguity and relative position to seller and buyer in the string. In cases of multiple

---

[145]Besides the two established data sources for M&A activities, I also exploit USPTO reassignments that have been classified as the result of a merger.

[146]The search engine can be accessed at http://www.bing.com [accessed: 22 July 2015].

and contradicting search results for the same owner buyer couples, the Bing ranking is used. I note that this method also allows a positive identification of arm's length patent transactions by finding the relevant keywords in the search results.

Geography

The MPIIC-PT2015, containing city, post code, and country code for all current and prior patent holders, allows the unambiguous identification of the location of the transacting parties. By utilizing the address information of current and prior rights holders, I first identify whether the transaction occurred on a *local, national* or *international* level. For the address, I first parsed the postcodes, city names, region names, and country codes into separate fields. I then updated the country codes to current jurisdictions, where necessary.[147] I harmonized cities according to available dictionaries. For city homonyms, I relied on the post code as further identification.

Since I mostly follow the address cleaning methodology outlined by Maraut *et al.* (2008), I assess the quality of the cleaning results by matching the data of this study via the raw address information with their publicly available data and find very few unambiguous discrepancies.[148]

---

[147]This applies primarily to countries of the former Soviet Union.
[148]The data are freely available at http://www.oecd.org/sti/inno/oecdpatentdatabases.htm [accessed: 22 July 2015].

Chapter 5

Summary

This thesis had two main topics: patent litigation and patent transfers. The first two studies addressed certain aspects of the German patent litigation system and their effect on litigant behavior. In the first study I examined what factors determine the patent holder's court selection given he has multiple courts available. I found that the determinants of court selection vary in magnitude depending on litigant and case characteristics. Expeditious patent enforcement is highly valued, if the litigants face each other in the same product market and the irrecoverable loss of rents due to delayed judgment is the largest. For large plaintiffs, the results show that the distance to a particular court has a relatively smaller negative effect on court selection. In contrast, small plaintiffs value local access to court more highly. There is also some evidence indicating that courts differ in their decision making. According to the results, there is considerable heterogeneity in the judges' tendency to grant a stay of proceeding. Furthermore, plaintiffs associate one of the courts, i.e., the regional court Munich, with a stronger anti-patentee bias relative to the other courts, and avoid it more likely if they could rely on prior experience with the court.

In the second study we showed theoretically as well as empirically that bifurcation as practiced in Germany strongly favors the patent holder in litigation. This occurs for two reasons: first, bifurcation creates a substantial number of cases where an invalid patent is held infringed, and second, fewer patents are challenged than we would expect compared to litigation systems where infringement and validity are dealt with jointly. The real possibility of facing an injunction based on an invalid patent may create legal uncertainty with implications on firm behavior. In fact, we observed that such legal uncertainty changes the opposition behavior of firms subject to a divergent decision: alleged infringers who experience an 'invalid but infringed' situation file more oppositions immediately afterwards.

In the third study I looked at a further way to appropriate rents from patents: patent transfers. Based on a set of arm's length patent transactions, I found evidence that there has been a considerable increase in activities on the European market for patents since the late 1990s. Nevertheless, the transfer rate by granted patents remains relatively constant at 7 to 8%. Compared to the transfer rates calculated by Serrano (2010) (13.5%) and Marco *et al.* (2015) (15.0%), the European market for patents appears underdeveloped. I also looked at the timing of patent grants relative to the events of patent transfer and observed a significant increase in the hazard of patent transfer after the internal communication of patent grant to the rights holder. This finding is in line with the notion that a decrease in uncertainty regarding a patent's validity and scope facilitates the success of negotiations between patent sellers and potential buyers.

One recurrent theme in all three studies represents the factor time. In Chapter 2, plaintiffs choose their court based on the expected time until judgment. In Chapter 3, the injunction gap, i.e., the time a patent can be legally enforced until it is subsequently invalidated, plays a crucial role regarding the incentives to challenge validity in the first place, and to engage in preemptive measures in the form of post-grant oppositions. Finally in Chapter 4, the hazard of patent transfer is considerably lower during the time of patent examination. The timing of decisions made by patent courts and patent offices evidently has a considerable impact on the overall efficiency of the patent system. While resource constraints obviously set boundaries to how fast high quality decisions on grant, infringement and validity can be made, the optimal institutional design of a patent system depends on the balance in the inherent tradeoff between scrutiny and promptness of decisionmaking to facilitate the enforcement and trade of patents, and ultimately to ensure incentives to innovate.

Bibliography

Adam, T. and Spence, M. (2001). Opposition in the European Patent Office: An Underestimated Weapon?, Oxford Intellectual Property Research Centre, Oxford.

Agrawal, A., Cockburn, I. and Zhang, L. (2015). Deals not Done: Sources of Failure in the Market for Ideas. *Strategic Management Journal*, 36 (7), 976–986.

Akcigit, U., Celik, M. A. and Greenwood, J. (2013). Buy, Keep or Sell: Economic Growth and the Market for Ideas, NBER Working Paper 19763.

Almeida, P. (1996). Knowledge Sourcing By Foreign Multinationals: Patent Citation Analysis in the U.S. Semiconductor Industry. *Strategic Management Journal*, 17 (2), 155–165.

— and Kogut, B. (1997). The Exploration of Technological Diversity and the Geographic Localization of Innovation. *Small Business Economics*, 9 (1), 21–31.

Andersson, D. E. (2014). How Does Stronger Patent Laws Affect Trade in Technology? Evidence from Patent Transfers in Sweden 1871-1914, Working Paper.

Ann, C. (2009). Verletzungsgerichtsbarkeit zentral für jedes Patentsystem und doch häufig unterschätzt. *GRUR – Gewerblicher Rechtsschutz und Urheberrecht*, 111 (3/4), 205–209.

— (2011). Technische Richter in der Patentgerichtsbarkeit – Ein Modell mit Perspektive? In *50 Jahre Bundespatentgericht - Festschrift zum 50-jährigen Bestehen des Bundespatentgerichts am 1. Juli 2011*, 111-127, Carl Heymanns Verlag.

—, Hauck, R. and Maute, L. (2011). *Auskunftsanspruch und Geheimnisschutz im Verletzungsprozess*. Carl Heymanns Verlag, 1st Edn.

Aoki, R. and Hu, J. (1999). Licensing vs. Litigation: The Effect of the Legal System on Incentives to Innovate. *Journal of Economics and Management Strategy*, 8 (1), 133–160.

— and — (2003). Time Factors of Patent Litigation and Licensing. *Journal of Institutional and Theoretical Economics*, 159 (2), 280–301.

Arora, A., Belenzon, S. and Rios, L. A. (2011). The Organization of R&D in American Corporations: The Determinants and Consequences of Decentralization, NBER Working Paper 17013.

— and Fosfuri, A. (2003). Licensing the Market for Technology. *Journal of Economic Behavior & Organization*, 52 (2), 277–295.

—, — and Gambardella, A. (2004). *Markets for Technology: The Economics of Innovation and Corporate Strategy*. MIT Press, 1st Edn.

— and Gambardella, A. (2010). The Market for Technology. In H. B. and N. Rosenberg (eds.), *Handbook of the Economics of Innovation*, Elsevier.

Atkinson, S. E., Marco, A. C. and Turner, J. L. (2009). The Economics of a Centralized Judiciary: Uniformity, Forum Shopping, and the Federal Circuit. *Journal of Law and Economics*, 52 (3), 411–443.

Ayres, I. and Klemperer, P. (1999). Limiting Patentees' Market Power Without Reducing Innovation Incentives: The Perverse Benefits of Uncertainty and Non-Injunctive Remedies. *Michigan Law Review*, 97 (4), 985–1033.

Bassett, D. L. (2006). The Forum Game. *North Carolina Law Review*, 84 (2), 333–396.

Battistella, C., Toni, A. F. D. and Pillon, R. (2015). Inter-organisational Technology/Knowledge Transfer: A Framework from Critical Literature Review. *Journal of Technology Transfer*, forthcoming.

Bebchuk, L. A. (1984). Litigation and Settlement under Imperfect Information. *The RAND Journal of Economics*, 15 (3), 404–415.

Bena, J. and Li, K. (2014). Corporate Innovations and Mergers and Acquisitions. *The Journal of Finance*, 69 (5), 1923–1960.

Bessen, J. and Meurer, M. J. (2008). *Patent Failure: How Judges, Bureaucrats, and Lawyers Put Innovators at Risk*. Princeton University Press, 1st Edn.

Bloom, N., Schankerman, M. and van Reenen, J. (2013). Identifying Technology Spillovers and Product Market Rivalry. *Econometrica*, 81 (4), 1347–1393.

Böhler, R. (2011). Einstweilige Verfügungen in Patentsachen. *GRUR – Gewerblicher Rechtsschutz und Urheberrecht*, 113 (11), 965–971.

Bösenberg, S. and Egger, P. (2014). R&D Tax Incentives and the Emergence and Trade of Ideas, Working Paper.

Boyce, J. R. and Hollis, A. (2007). Preliminary Injunctions and Damage Rules in Patent Law. *Journal of Economics and Management Strategy*, 16 (2), 385–405.

Bozeman, B. (2000). Technology Transfer and Public Policy: A Review of Research and Theory. *Research Policy*, 29 (4-5), 627–655.

Burhop, C. (2010). The Transfer of Patents in Imperial Germany. *The Journal of Economic History*, 70 (4), 921–939.

— and Wolf, N. (2013). The German Market for Patents during the "Second Industrialization," 1884-1913: A Gravity Approach. *Business History Review*, 87 (1), 63–69.

Burke, P. F. and Reitzig, M. (2007). Measuring Patent Assessment Quality – Analyzing the Degree and Kind of (In)consistency in Patent Offices' Decision Making. *Research Policy*, 36 (9), 1404–1430.

Cabrillo, F. and Fitzpatrick, S. (2008). *The Economics of Courts and Litigation*. Edward Elgar Publishing, 1st Edn.

Callaert, J., Plessis, M. D., Grouwels, J., Lecocq, C., Magerman, T., Peeters, B., Song, X., Looy, B. V. and Vereyen, C. (2011). Patent Statistics at Eurostat: Methods for Regionalisation, Sector Allocation and Name Harmonisation, Eurostat Methodologies and Working Papers.

—, Van Looy, B., Verbeek, A., Debackere, K. and Thijs, B. (2006). Traces of Prior Art: An Analysis of Non-patent References Found in Patent Documents. *Scientometrics*, 69 (1), 3–20.

Carpenter, M. P., Narin, F. and Woolf, P. (1981). Citation Rates to Technologically Important Patents. *World Patent Information*, 3 (4), 160–163.

Carpenter, R. and Petersen, B. (2002). Is the Growth of Small Firms Constrained by Internal Finance? *The Review of Economics and Statistics*, 84 (2), 298–309.

Cassiman, B., Veugelers, R. and Zuniga, P. (2008). In Search of Performance Effects of (In)direct Industry Science Links. *Industrial and Corporate Change*, 17 (4), 611–646.

Caviggioli, F. and Ughetto, E. (2013). The Drivers of Patent Transactions: Corporate Views on the Market for Patents. *R&D Management*, 43 (4), 318–332.

Chesbrough, H. (2006). Emerging Secondary Markets for Intellectual Property: US and Japan Comparisons.

Chien, C. V. (2010). From Arms Race to Marketplace: The Complex Patent Ecosystem and Its Implications for the Patent System. *Hastings Law Journal*, 62 (1), 297–356.

— and Helmers, C. (2015). Inter Partes Review and the Design of Post-Grant Patent Reviews. *Stanford Technology Law Review, forthcoming*, forthcoming.

Choi, J. P. and Gerlach, H. A. (2013). A Theory of Patent Portfolios, CESifo Working Paper Series No. 4405.

Cohen, W. M., Nelson, R. R. and Walsh, J. P. (2000). Protecting Their Intellectual Property Assets: Appropriability Conditions and Why U.S. Manufacturing Firms Patent (or Not), NBER Working Paper 7552.

Conti, R., Gambardella, A. and Novelli, E. (2013). Research on Markets for Inventions and Implications for R&D Allocation Strategies. *The Academy of Management Annals*, 7 (1), 717–774.

Cotropia, C. A., Kesan, J. P. and Schwartz, D. L. (2014). Unpacking Patent Assertion Entities (PAEs). *Minnesota Law Review*, 99 (2), 649–704.

Cotter, T. F. (2013). *Comparative Patent Remedies: A Legal and Economic Analysis*. Oxford University Press, Oxford, 1st Edn.

Cremers, K., Ernicke, M., Gaessler, F., Harhoff, D., Helmers, C., McDonagh, L., Schliessler, P. and van Zeebroeck, N. (2013). Patent Litigation in Europe, Center for European Economic Research Discussion Paper No. 13-072.

—, Gaessler, F., Harhoff, D. and Helmers, C. (2014). Invalid but Infringed? An Analysis of Germany's Bifurcated Patent Litigation System, Max Planck Institute for Innovation & Competition Research Paper No. 14-14.

Czarnitzki, D., Hussinger, K. and Schneider, C. (2015). R&D Collaboration with Uncertain Intellectual Property Rights. *Review of Industrial Organization*, 46 (2), 183–204.

Daughety, A. F. and Reinganum, J. F. (2012). Settlement. In *The Encyclopedia of Law and Economics – Procedural Law and Economics*, Vol. 8, 2nd Edn., Edward Elgar Publishing, 386–471.

Drivas, K. and Economidou, C. (2015). Is Geographic Nearness Important for Trading Ideas? Evidence from the US. *The Journal of Technology Transfer*, 40 (4), 629–662.

Dyer, J. H. and Singh, H. (1998). The Relational View: Cooperative Strategy and Sources of Interorganizational Competitive Advantage. *The Academy of Management Review*, 23 (4), 660–679.

EPO-OHIM (2013). *Intellectual Property Rights Intensive Industries: Contribution to Economic Performance and Employment in the European Union*. Tech. rep., European Patent Office and the Offie for Harmonization in the Internal Market.

Farrell, J. and Merges, R. P. (2004). Incentives to Challenge and Defend Patents: Why Litigation Won't Reliably Fix Patent Office Errors and Why Administrative Patent Review Might Help. *Berkeley Technology Law Journal*, 19 (3), 943–970.

Feldman, M. P. and Kogler, D. F. (2010). *Handbook of the Economics of Innovation*, University of Chicago Press, Vol. 1, Chap. Stylized Facts in the Geography of Innovation, 381–410. 1st Edn.

Fischer, T. and Henkel, J. (2012). Patent Trolls on Markets for Technology – An Empirical Analysis of NPEs' Patent Acquisitions. *Research Policy*, 41, 1519–1533.

— and Leidinger, J. (2014). Testing Patent Value Indicators on Directly Observed Patent Value – An Empirical Analysis of Ocean Tomo Patent Auctions. *Research Policy*, 43, 519–529.

Fock, S. and Bartenbach, K. (2010). Zur Aussetzung nach § 148 ZPO bei Patentverletzungsverfahren. *Mitteilungen der deutschen Patentanwälte*, 101 (4), 155–161.

Galasso, A. and Schankerman, M. (2010). Patent Thickets, Courts, and the Market for Innovation. *The RAND Journal of Economics*, 41 (3), 472–503.

—, — and Serrano, C. J. (2013). Trading and Enforcing Patent Rights. *The RAND Journal of Economics*, 44 (2), 275–312.

Gambardella, A. and Giarratana, M. S. (2013). General Technological Capabilities, Product Market Fragmentation, and Markets for Technology. *Research Policy*, 42 (2), 315–325.

—, Giuri, P. and Luzzi, A. (2007). The Market for Patents in Europe. *Research Policy*, 36 (8), 1163–1183.

Gans, J. S., Hsu, D. H. and Stern, S. (2008). The Impact of Uncertain Intellectual Property Rights on the Market for Ideas: Evidence from Patent Grant Delays. *Management Science*, 54 (5), 982–997.

Gilbert, R. and Shapiro, C. (1990). Optimal Patent Length and Breadth. *The RAND Journal of Economics*, 21 (1), 106–112.

Graham, S. and van Zeebroeck, N. (2014). Comparing Patent Litigation Across Europe: A First Look. *Stanford Technology Law Review*, 17 (3), 655–708.

Griffith, R., Miller, H. and O'Connell, M. (2014). Ownership of Intellectual Property and Corporate Taxation. *Journal of Public Economics*, 112 (1), 12–23.

Grimpe, C. and Hussinger, K. (2008). Pre-empting Technology Competition Through Firm Acquisitions. *Economics Letters*, 100 (2), 189–191.

Guellec, D. and van Pottelsberghe de la Potterie, B. (2000). Applications, Grants and the Value of Patent. *Economics Letters*, 69 (1), 109–114.

Hagiu, A. and Yoffe, D. B. (2013). The New Patent Intermediaries: Platforms, Defensive Aggregators, and Super-Aggregators. *Journal of Economic Perspectives*, 27 (1), 45–66.

Hall, B., Jaffe, A. and Trajtenberg, M. (2001). The NBER Patent Citations Data File: Lessons, Insights, and Methodological Tools, NBER Working Paper. 8498.

Hall, B. H. (2002). The Financing of Research and Development. *Oxford Review of Economic Policy*, 18 (1), 35–51.

— and Ziedonis, R. H. (2001). The Patent Paradox Revisited: An Empirical Study of Patenting in the U.S. Semiconductor Industry, 1979-1995. *The RAND Journal of Economics*, 32 (1), 101–128.

Harguth, A. and Carlson, S. C. (2011). *Patents in Germany and Europe, Procurement, Enforcement and Defence: An International Handbook*. Kluwer Law International, 1st Edn.

Harhoff, D. (2005). The Battle for Patent Rights. In M. Mejer and B. van Pottelsberghe de la Potterie (eds.), *Economics and Management Perspectives on Intellectual Property Rights*, 21–39, Palgrave-McMillan.

—, Hoisl, K., Reichl, B. and van Pottelsberghe de la Potterie, B. (2009). Patent Validation at the Country Level – The Role of Fees and Translation Costs. *Research Policy*, 38 (9), 1423–1437.

— and Reitzig, M. (2004). Determinants of Opposition against EPO Patent Grants – The Case of Biotechnology and Pharmaceuticals. *International Journal of Industrial Organization*, 22 (4), 443–480.

—, Scherer, F. M. and Vopel, K. (1999). Citation Frequency and the Value of Patented Inventions. *The Review of Economics and Statistics*, 81 (3), 511–515.

—, — and — (2003). Citations, Family Size, Opposition and the Value of Patent Rights. *Research Policy*, 32 (8), 1343–1363.

— and Wagner, S. (2009). The Duration of Patent Examination at the European Patent Office. *Management Science*, 55 (12), 1969–1984.

Helmers, C., McDonagh, L. and Love, B. (2014). Is There a Patent Troll Problem in the UK? *Fordham Intellectual Property, Media & Entertainment Law Journal*, 24, 509–553.

Henry, M. D. and Turner, J. L. (2006). The Court of Appeals for the Federal Circuit's Impact on Patent Litigation. *Journal of Legal Studies*, 35, 85–118.

Herr, J. and Grunwald, M. (2011). Speedy Patent Infringement Proceedings in Germany: Pros and Cons of the Go-To Courts. *Journal of Intellectual Property Law & Practice*, 7 (1), 44–47.

Hilty, R. M., Jaeger, T., Lamping, M. and Ullrich, H. (2012). The Unitary Patent Package: Twelve Reasons for Concern. *Max Planck Institute for Intellectual Property and Competition Law*.

— and Lamping, M. (2011). Trennungsprinzip – Quo vadis, Germania? In *50 Jahre Bundespatentgericht - Festschrift zum 50-jährigen Bestehen des Bundespatentgerichts am 1. Juli 2011*, 255–273, Carl Heymanns Verlag.

Hochberg, Y. V., Serrano, C. J. and Ziedonis, R. H. (2015). Patent Collateral, Investor Commitment, and the Market for Venture Lending, Working Paper.

Jensen, P. H., Palangkaraya, A. and Webster, E. (2015). Trust and the Market for Technology. *Research Policy*, 44 (2), 340–356.

Kaess, T. (2009). Die Schutzfähigkeit technischer Schutzrechte im Verletzungsverfahren. *GRUR – Gewerblicher Rechtsschutz und Urheberrecht*, 111, 276–281.

Karkinsky, T. and Riedel, N. (2012). Corporate Taxation and the Choice of Patent Location within Multinational Firms. *Journal of International Economics*, 88 (1), 176–185.

Kesan, J. P. and Ball, G. G. (2011). Judicial Experience and the Efficiency and Accuracy of Patent Adjudication: An Empirical Analysis of the Case for a Specialized Patent Trial Court. *Harvard Journal of Law & Technology*, 24 (2), 393–467.

Keukenschrijver, A. (1999). *Die gerichtliche Durchsetzung von technischen Schutzrechten*, *Freiberger Seminare zum Gewerlichen Rechtsschutz*, Vol. 1.

Khan, B. Z. and Sokoloff, K. L. (2004). Institutions and Democratic Invention in 19th-Century America: Evidence from "Great Inventors," 1790-1930. *The American Economic Review*, 94 (2), 395–401.

Klerman, D. M. and Reilly, G. (2014). Forum Selling, USC Law Legal Studies Paper No. 14-44.

Kraßer, R. (2009). *Patentrecht – Ein Lehr- und Handbuch zum deutschen Patent- und Gebrauchsmusterrecht, Europäischen und Internationalen Patentrecht*. C. H. Beck München, 6th Edn.

Kühnen, T. (2009). What Becomes of Judgments on Infringement when the Patents in Suit are Revoked: the Legal Situation in Germany. In *Special edition 1/2009 14th European Patent Judges' Symposium*, 56–63, European Patent Office.

Kühnen, T. (2012). *Patent Litigation Proceedings in Germany: A Handbook for Practitioners*. translated by Frank Peterreins, Carl Heymanns Verlag, 6th Edn.

Kühnen, T. (2013). The Bifurcation System in German Practice. In *Special edition 1/2013 16th European Patent Judges' Symposium*, 59–93, European Patent Office.

Lamoreaux, N. R. and Sokoloff, K. L. (1999a). Inventive Activity and the Market for Technology in the United States, 1840-1920, NBER Working Paper 7107.

— and — (1999b). *Inventors, Firms, and the Market for Technology in the Late Nineteenth and Early Twentieth Centuries*, University of Chicago Press, Chap. 1, 19–60.

— and — (2001). Market Trade in Patents and the Rise of a Class of Specialized Inventors in the 19th-Century United States. *The American Economic Review*, 91 (2), 39–44.

Lanjouw, J. O. and Lerner, J. (2001). Tilting the Table? The Use of Preliminary Injunctions. *Journal of Law and Economics*, 44 (2), 573–603.

—, Pakes, A. and Putnam, J. (1998). How to Count Patents and Value Intellectual Property: The Uses of Patent Renewal and Application Data. *The Journal of Industrial Economics*, 46 (4), 405–432.

— and Schankerman, M. (2001). Characteristics of Patent Litigation: A Window on Competition. *The RAND Journal of Economics*, 32 (1), 129–151.

Lemley, M. A. (2001). Rational Ignorance at the Patent Office. *Northwestern University Law Review*, 95 (4), 1–34.

— (2010). Where to File Your Patent Case. *AIPLA Quarterly Journal*, 38 (4), 401–436.

—, Kendall, J. and Martin, C. (2013). Rush to Judgment? Trial Length and Outcomes in Patent Cases. *AIPLA Quarterly Journal*, 41 (2), 169–204.

— and Shapiro, C. (2005). Probabilistic Patents. *Journal of Economic Perspectives*, 19 (2), 75–98.

Lerner, J. (1994). The Importance of Patent Scope: An Empirical Analysis. *The RAND Journal of Economics*, 25 (2), 319–333.

— and Seru, A. (2015). The Use and Misuse of Patent Data: Issues for Corporate Finance and Beyond, Working Paper.

— and Tirole, J. (2006). A Model of Forum Shopping. *The American Economic Review*, 96 (4), 1091–1113.

Lii, T. (2013). Shopping for Reversals: How Accuracy Differs across Patent Litigation Forums. *The Chicago-Kent Journal of Intellectual Property*, 12, 31–51.

Love, B. J. (2009). The Misuse of Reasonable Royalty Damages as a Patent Infringement Deterrent. *Missouri Law Review*, 74 (4), 909–948.

Magerman, T., Grouwels, J., Song, X. and Looy, B. V. (2009). Data Production Methods for Harmonized Patent Indicators: PaPatent Name Harmonization, EUROSTAT Working Paper and Studies, Luxembourg.

Mann, R. J. and Underweiser, M. (2012). A New Look at Patent Quality: Relating Patent Prosecution to Validity. *Journal of Empirical Legal Studies*, 9 (1), 1–32.

Maraut, S., Dernis, H., Webb, C., Spiezia, V. and Guellec, D. (2008). *The OECD Regpat Database: A Presentation*. Tech. rep., OECD STI Working Paper 2008/2 – Statistical Analysis of Science, Technology and Industry.

Marco, A. C., Myers, A. F., Graham, S., D'Agostino, P. and Apple, K. (2015). The USPTO Patent Assignment Dataset: Descriptions, Lessons, and Insights, USPTO Working Paper No. 20150X.

Martínez, C. (2011). Licensing and Change of Ownership in International Patent Legal Status Data. http://www.oecd.org/sti/sci-tech/49363502.pdf, [accessed: 22 July 2015].

McKelvie, R. R. (2007). Forum Selection in Patent Litigation: A Traffic Report. *Intellectual Property & Technology Law Journal*, 19 (8), 1–17.

Mejer, M. and van Pottelsberghe de la Potterie, B. (2011). Patent Backlogs at USPTO and EPO: Systemic Failure vs Deliberate Delays. *World Patent Information*, 33 (2), 122–127.

— and — (2012). Economic Incongruities in the European Patent System. *European Journal of Law and Economics*, 34 (1), 215–234.

Menell, P. S. (2000). Intellectual Property: General Theories. In B. Bouckaert and G. de Geest (eds.), *Encyclopedia of Law & Economics*, Vol. 2, *1600*, Edward Elgar, Cheltenham, 129–188.

Ménière, Y. and Dechezleprêtre, A. (2012). The Market for Patents in Europe 1997-2009. http://www.cerna.mines-paristech.fr/images/stories/Menire/MENIERE/YM%20PSDM.pdf, [accessed: 22 July 2015].

Merges, R. P. (1994). Of Property Rules, Coase, and Intellectual Property. *Columbia Law Review*, 94 (8), 2655–2673.

— (1999). As Many as Six Impossible Patent before Breakfast: Property Rights for Business Concepts and Patent System Reform. *Berkeley Technology Law Journal*, 14 (2), 577–616.

Monk, A. H. B. (2009). The Emerging Market for Intellectual Property: Drivers, Restrainers, and Implications. *Journal of Economic Geography*, 9 (4), 469–491.

Moore, K. A. (2001a). Are District Court Judges Equipped to Resolve Patent Cases? *Harvard Journal of Law & Technology*, 15 (1), 1–39.

— (2001b). Forum Shopping in Patent Cases: Does Geographic Choice Affect Innovation? *North Carolina Law Review*, 79, 889–938.

Müller-Stoy, T. and Schachl, T. (2011). LG München I macht Lizenzanalogie attraktiver. *GRUR-Prax – Gewerblicher Rechtsschutz und Urheberrecht, Praxis im Immaterialgüter- und Wettbewerbsrecht*, 3, 341–343.

Müller-Stoy, T. and Wahl, J. (2008). Düsseldorfer Praxis zur einstweiligen Unterlassungsverfügung wegen Patentverletzung. *Mitteilungen der deutschen Patentanwälte*, 99 (78), 311–313.

Münster-Horstkotte, A. (2012). Das Trennungsprinzip im deutschen Patentsystem – Probleme und Lösungsmöglichkeiten. *Mitteilungen der deutschen Patentanwälte*, 103 (1), 1–9.

Nicholas, T. and Shimizu, H. (2013). Intermediary Functions and the Market for Innovation in Meiji and Taishō Japan. *Business History Review*, 87 (1), 121–149.

Nordhaus, W. D. (1969). *Invention, Growth and Welfare: A Theoretical Treatment of Technological Change*. M.I.T. Press.

Novelli, E. (2015). An Examination of the Antecedents and Implications of Patent Scope. *Research Policy*, 44 (2), 493–507.

Odasso, C., Scellato, G. and Ughetto, E. (2015). Selling Patents at Auction: An Empirical Analysis of Patent Value. *Industrial and Corporate Change*, 24 (2), 417–438.

Oropo (2015). Who Owns the World's Patents? http://oropo.net/oropo_report_20150615.pdf [accessed: 29 June 2015].

Ozimek, A. and Miles, D. (2011). Stata Utilities for Geocoding and Generating Travel Time and Travel Distance Information. *The Stata Journal*, 11 (1), 106–119.

Palermo, V., Higgins, M. J. and Ceccagnoli, M. (2015). Assets with "Warts": How Reliable is the Market for Technology?, NBER Working Paper 21103.

Parhankangas, A. and Arenius, P. (2003). From a Corporate Venture to an Independent Company: A Base for a Taxonomy for Corporate Spin-off Firms. *Research Policy*, 32 (3), 463–481.

Pegram, J. B. (2000). Should There Be a U.S. Trial Court With a Specialization in Patent Litigation? *Journal of the Patent and Trademark Office Society*, 82 (2), 765–796.

Pitz, J. (2011). Entscheidungsharmonie in Patentstreitverfahren. In *50 Jahre Bundespatentgericht – Festschrift zum 50-jährigen Bestehen des Bundespatentgerichts am 1. Juli 2011*, 433–447, Carl Heymanns Verlag.

Posner, R. A. (1993). What Do Judges and Justices Maximize? (The Same Thing Everybody Else Does). *Supreme Court Economic Review*, 3, 1–41.

Powell, W. W., Koput, K. W. and Smith-Doerr, L. (1996). Interorganizational Collaboration and the Locus of Innovation: Networks of Learning in Biotechnology. *Administrative Science Quarterly*, 41 (1), 116–145.

Raffo, J. and Lhuillery, S. (2009). How to Play the "Names Game": Patent Retrieval Comparing Different Heuristics. *Research Policy*, 38 (6), 1617–1627.

Régibeau, P. and Rockett, K. (2010). Innovation Cycles and Learning at the Patent Office: Does the Early Patent Get the Delay? *The Journal of Industrial Economics*, 58 (2), 222–246.

Scellato, G., Calderini, M., Caviggioli, F., Franzoni, C., Ughetto, E., Kica, E. and Rodriguez, V. (2011). Study on the Quality of the Patent System in Europe, Final Report Tender No. MARKT/2009/11/D.

Schankerman, M. (1998). How Valuable is Patent Protection? Estimates by Technology Field. *The RAND Journal of Economics*, 29 (1), 77–107.

Schmoch, U. (2008). Concept of a Technology Classification for Country Comparisons, Final Report to the World Intellectual Property Organisation (WIPO).

Schramm, C. and Kaess, T. (2010). *Der Patentverletzungsprozess*. Carl Heymanns Verlag, 6th Edn.

Scotchmer, S. (1991). Standing on the Shoulders of Giants: Cumulative Research and the Patent Law. *The Journal of Economic Perspectives*, 5 (1), 29–41.

Serrano, C. J. (2010). The Dynamics of the Transfer and Renewal of Patents. *The RAND Journal of Economics*, 41 (4), 686–708.

— (2013). Estimating the Gains from Trade in the Market for Patent Rights.

Shapiro, C. (2001). Navigating the Patent Thicket: Cross Licensing, Patent Pools, and Standard Setting. In A. B. Jaffe, J. Lerner and S. Stern (eds.), *Innovation Policy and the Economy*, Vol. 1, M.I.T. Press, 119–150.

Sneed, K. A. and Johnson, D. K. N. (2009). Selling Ideas: The Determinants of Patent Value in An Auction Environment. *R&D Management*, 39 (1), 87–94.

Spier, K. E. (2007). Litigation. In *The Handbook of Law and Economics*, 4, 1st Edn., North Holland.

Spulber, D. F. (2015). How Patents Provide the Foundation of the Market for Inventions. *Journal of Competition Law & Economic*, forthcoming.

Stauder, D. (1983). Die tatsächliche Bedeutung von Verletzungs- und Nichtigkeitsverfahren in der Bundesrepublik Deutschland, Frankreich, Großbritannien und Italien – Ergebnisse einer statistisch-empirischen Untersuchung. *GRUR International – Gewerblicher Rechtsschutz und Urheberrecht, Internationaler Teil*, 32, 233–242.

— (1989). *Patent- und Gebrauchsmusterverletzungsverfahren in der Bundesrepublik Deutschland, Großbritannien, Frankreich und Italien – Eine rechtstatsächliche Untersuchung.* Carl Heymanns Verlag, 1st Edn.

Stephenson, M. C. (2009). Legal Realism for Economists. *Journal of Economic Perspectives*, 23 (2), 191–211.

Stieger, W. (2004). EIPIN Patent Litigation Congress – Pre-Litigation Strategies. *IIC – International Review of Intellectual Property and Competition Law*, 35 (5), 551–559.

Thoma, G., Torrisi, S., Gambardella, A., Guellec, D., Hall, B. H. and Harhoff, D. (2010). Harmonizing and Combining Large Datasets – An Application to Firm-Level Patent and Accounting Data, NBER Working Paper No. 15851.

Thomas, J. R. (2002). The Responsibility of the Rulemaker: Comparative Approaches to Patent Administration Reform. *Berkeley Technology Law Journal*, 17, 727–761.

Trajtenberg, M. (1990). A Penny for Your Quotes: Patent Citations and the Value of Innovation. *The RAND Journal of Economics*, 21 (1), 172–187.

Ullrich, H. (2015). The European Patent and Its Courts: An Uncertain Prospect and an Unfinished Agenda. *IIC - International Review of Intellectual Property and Competition Law*, 46 (1), 1–9.

Van de Ven, W. P. M. M. and Van Praag, B. M. (1981). The Demand for Deductibles in Private Health Insurance: A Probit Model with Sample Selection. *Journal of Econometrics*, 17 (2), 229–252.

Van Pottelsberghe de la Potterie, B. (2015). *Encyclopedia of Law and Economics*, Springer New York, Chap. European Patent System, 1–9.

— and van Zeebroeck, N. (2008). A Brief History of Space and Time: The Scope-year Index as a Patent Value Indicator Based on Families and Renewals. *Scientometrics*, 75, 319–338.

van Zeebroeck, N., Stevnsborg, N., van Pottelsberghe de la Potterie, B., Guellec, D. and Archontopoulos, E. (2008). Patent Inflation in Europe. *World Patent Information*, 30 (1), 43–52.

Vandermeulen, B. (2005). Harmonization of IP Litigation Practice – Still a Long Road Ahead. *Journal of Intellectual Property Law & Practice*, 1 (1), 30–36.

Veugelers, R. and Cassiman, B. (1999). Make and Buy in Innovation Strategies: Evidence from Belgian Manufacturing Firms. *Research Policy*, 28 (1), 63–80.

von Hees, A. and Braitmayer, S.-E. (2010). *Verfahrensrecht in Patentsachen*. Carl Heymanns Verlag, 4th Edn.

Wadlow, C. (2015). An Historical Perspective II: The Unified Patent Court. In J. Pila and C. Wadlow (eds.), *The Unitary EU Patent System*, 3, 1st Edn., Hart Publishing, Oxford, 33–44.

Wang, A. W. (2010). Rise of the Patent Intermediaries. *Berkeley Technology Law Journal*, 25, 159–200.

Weatherall, K. and Webster, E. (2014). Patent Enforcement: A Review of the Literature. *Journal of Economic Surveys*, 28 (2), 312–343.

Williamson, O. E. (1985). *The Economic Institutions of Capitalism*. The Free Press, New York, 1st Edn.

— (1991). Comparative Economic Organization: The Analysis of Discrete Structural Alternatives. *Administrative Science Quarterly*, 36 (2), 269–296.

Yanagisawa, T. and Guellec, D. (2009). The Emerging Patent Marketplace, STI Working Paper 2009/9, OECD.

Ziedonis, R. H. (2004). Don't Fence Me in: Fragmented Markets for Technology and the Patent Acquisition Strategies of Firms. *Management Science*, 50 (6), 804–820.